# Mysteries of the Milky Way

DONALD GOLDSMITH
and NATHAN COHEN



CB
CONTEMPORARY
BOOKS
CHICAGO

**Library of Congress Cataloging-in-Publication Data**

Goldsmith, Donald.
Mysteries of the Milky Way / Donald Goldsmith and Nathan Cohen.
p. cm.
ISBN 0-8092-4189-7 (cloth)
ISBN 0-8092-3932-9 (paper)
1. Milky Way. I. Cohen, Nathan. II. Title.
QB857.7.G65 1991
523.1'13—dc20 90-28084
CIP

Front cover photograph: This star-forming region lies thousands of light-years from the solar system in the direction of the constellation Carina.


Published by Contemporary Books, Inc.
180 North Michigan Avenue, Chicago, Illinois 60601
Manufactured in the United States of America
International Standard Book Number: 0-8092-4189-7 (cloth)
0-8092-3932-9 (paper)

To
Rachel Goldsmith
and
Mary Louise Shelman

# Contents

# 1
# THE MILKY WAY AND THE BOWL OF NIGHT

GO OUTSIDE ON a clear night and look at the sky: What do you see? For many human generations, the most striking feature of the sky on a moonless night was the "milky way," the streak of light that stretches from horizon to horizon. This shimmering band of undifferentiated light, stretched and distorted by lanes of darkness that add to its form and definition, seems quite different in nature from the points of light we call stars. Quite naturally, the seventeenth-century poet John Suckling, living in an age before artificial lighting had appeared, wrote of a woman whose "face is like the milky way i' the sky—a meeting of gentle lights without a name."

The opportunity to admire the milky way was one of the first sacrifices that humanity made in adapting to advanced city life. Once large numbers of people began to light their homes and their streets at night, they produced artificial light that is scattered by the air. This scattered light interferes enormously with our ability to see faint objects in the heavens. Today, even in Los Angeles or Tokyo, you can see the brightest stars on a clear night—but never the milky way.

So take yourself to the country and turn your gaze to the heavens on a clear moonless night. Better yet, forget everything you once knew about the milky way—everything you might read in a book; forget, in fact, that there *is* a cosmos beyond the Earth, and then, just when you have no thought whatsoever of examining the night sky, the milky way will leap out of nowhere, stunning you with its brightness. The broad band of light—subtle but unmistakable—that stretches from horizon to horizon, reaching halfway up to the zenith, will be something you've never seen before—indeed, something that you may have never thought you *could* see before. You might even decide that you have had a deeper experience than those who live, or lived, where no city lights could interfere with the heavenly views: Those poor souls have had the milky way before them on every clear night since they could notice it, whereas you have saved this peak experience, almost without noticing that you were doing so, for the appropriate moment when you could truly appreciate what nature has spread around you.

What then is this milky way, the French *voie lactee*, the Russian *mlechniy put'*? You are seeing the disklike galaxy in which our solar system exists—seeing it from the inside, so that you look outward through the disk that appears to circle the Earth (see Figure 1). Of course, half of the milky way at any given time is below the horizon, but if the Earth were transparent and the sun nonexistent, you would note that the milky way does form a complete circle. Starting in the constellation Sagittarius (for reasons that will eventually become clear), the milky way runs

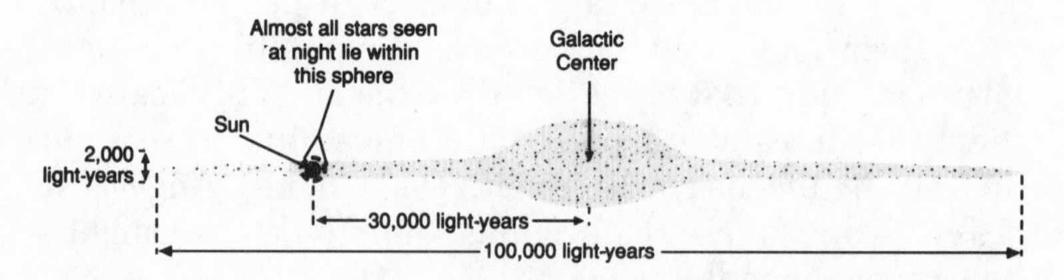


Figure 1: Because the solar system lies within the disk-shaped Milky Way galaxy, we see the "milky way" in the sky when we look in directions along the disk but not when we look outward from the plane of the galaxy. Nearly all the stars seen with the naked eye lie within a few hundred light-years of Earth—a region that spans only 1 percent or so of the Milky Way. *Drawing by Marjorie Baird Garlin*

westward through the fishhook tail of Scorpius, then southward through Centaurus, Crux, Carina, and Vela, up north again through the dim constellation of Puppis, toward Canis Major and Orion, and north farther still through Auriga, Perseus, Cassiopeia, and Cepheus, then to the south again via Cygnus and Aquila, and on through Scutum before joining Sagittarius once again (see Figure 2).

In all these directions, myriad stars of our galaxy, invisible as individuals, crowd together to create a milky band of light. The ancient Greeks apparently named it first: The Greek word *galaxos* means milky. *Galaxos* gave us our word *galaxy* along with the concept of a milky way in the sky. Two millennia later, when Galileo turned his crude telescope toward the milky way in 1609, he observed the diffuse light resolved into a host of bright points—clearly stars dimmer than could be seen individually by the un-

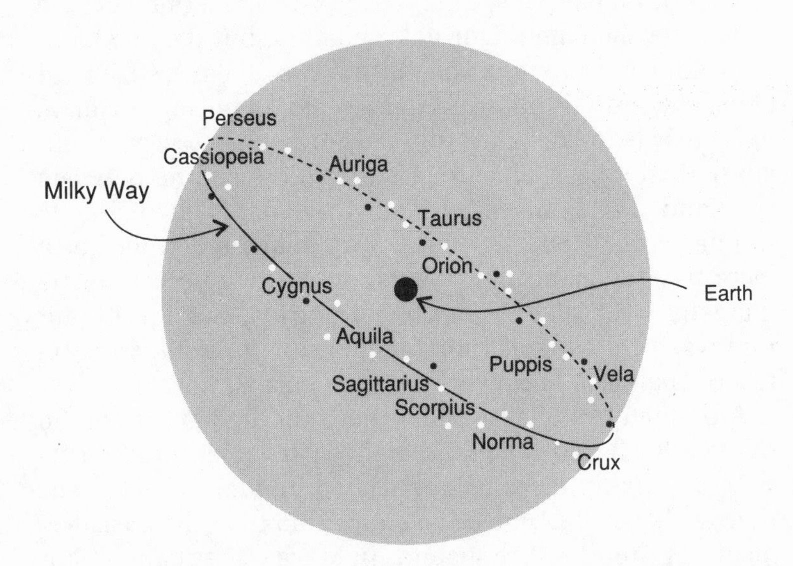


Figure 2: On the imaginary celestial sphere that holds all the stars at the same distance from Earth, the plane of the Milky Way passes through a band of constellations that includes Perseus, Cygnus, Aquila, and Sagittarius, which contains the center of our galaxy. *Drawing by Crystal Stevenson*

aided eye. Only then did humans acquire a basic understanding of what caused the band to appear across the sky (see Photo 1). Another three centuries sufficed to make it clear that the milky way in the sky derives its shape from the fact that we live in an "island universe." Our galaxy has a flattened, disklike shape that causes us to see the bulk of its stars spread around the sky. Today we call our galaxy the Milky Way, using the capital letters for the object, and the lowercase milky way for the band of light that we observe on Earth—the Milky Way as seen from our particular vantage point.

Once astronomers realized during the first decades of this century that we live *inside* a galaxy, they came to the related conclusion that other galaxies exist—indeed, exist by the billions—in the vast cosmos that we call home. Until times well within the lifetimes of living men and women, astronomers believed the Milky Way to be the largest of all galaxies. Eventually astronomers reached a different conclusion: Our galaxy is big, but it's no champion among galaxies. The Milky Way measured out at about a hundred thousand light-years from rim to rim. A light-year is the distance that light travels in a year, about six trillion miles; the closest stars to the sun lie between four and five light-years from the solar system. In its thinner dimension, the Milky Way spans a distance of a mere few thousand light-years, only a few percent of its diameter: Our galaxy, like most spirals, has an elegant thinness to its shape, more comparable to a phonograph record than to a hockey puck (see Photo 2).

A diameter of one hundred thousand light-years and a thickness of five thousand light-years does provide us with a suitably large neighborhood, though we shall see that we haven't included the entire galaxy in the measurement. A hundred thousand light-years equals about twenty thousand times the distance to the sun's closest neighbors and about two hundred times the distance to the farthest star you can see with the naked eye. Therefore, when the milky way catches your attention for the first time as a backdrop to the thousands of stars on the night sky, you should not fail to note that all these stars crowd into the merest fraction of the galaxy that surrounds us. The solar system lies about thirty thousand light-years from the center of the Milky Way. (These distances, like most others said to be "well estimated" in our galaxy, are accurate to plus or minus 10 or 20 percent.) The sun lies within two hundred light-years of the Milky Way's plane of symmetry, the imaginary plane that divides the galaxy into a top half and a bottom half, fundamentally identical. The stars of night all lie within five hundred light-years of that plane, as do nearly all the stars in the Milky Way. No star visible to the unaided eye ranks among the "outriders" of our galaxy, the stars that lie much more than five hundred light-years above or below the galactic plane, straggling outward to merge with intergalactic space. All in all, our naked-eye view of the stars at night encompasses less than one-hundredth of a percent of the total volume of the galaxy in which we live (see Figure 1).

This statement must be qualified by two exceptions. First, the stars of the Milky Way itself lie much farther from us than all the stars we see as individual objects. Toward Sagittarius and Scorpius, in the direction of the galactic center, we look almost thirty thousand light-years—a thousand times farther than most of the stars we see as points—to see the stars that crowd together in the band of light. In other directions—for example, when we look in the direction opposite to the center, in Perseus and Cassiopeia—we can also see stars tens of thousands of light-years away, likewise hundreds of times more distant than the stars in Orion or Scorpius.

The second exception to the rule that we see only our most immediate neighbors is the *galaxies* we can see, most notably the Andromeda galaxy, which lies to the west of the two chains of Andromeda, above the great square of Pegasus (see Figure 3). Shining with light that has taken two million years to reach us, this galaxy appears behind a backdrop of stars in the Milky Way, those stars that

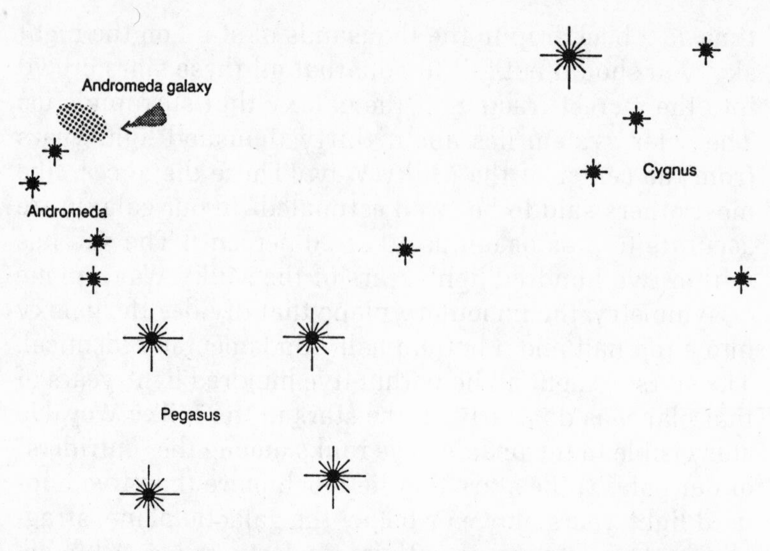


Figure 3: To find the Andromeda galaxy, look during the summer or fall for the constellation Pegasus. Then use the stars of Andromeda, which joins Pegasus at one corner, to find the Andromeda spiral, which will appear only as a faint, diffuse mass of light. *Drawing by Crystal Stevenson*

happen to lie just in the direction of our neighbor galaxy (see Photo 3). None of these stars is bright enough to be seen without a telescope, and the average distance to these stars is about a thousand light-years—one two-thousandth of the distance to Andromeda.

The Andromeda galaxy offers us a chance to look at our own Milky Way—not at our galaxy itself but at a close duplicate, so far as astronomers can determine. Like the Milky Way, the Andromeda galaxy consists of a disk of stars a hundred thousand light-years in diameter. Within that disk, the youngest and brightest stars appear in "spiral arms." The spiral arms are patterns that reflect the regions of most recent star birth. Highly luminous stars last only a few million years before they burn themselves out and explode as supernovae in one last glorious gasp. Therefore, when we see such young, luminous stars, we see the record of astronomically recent star birth. A spiral galaxy's pinwheel-like spiral arms arise from the fact that

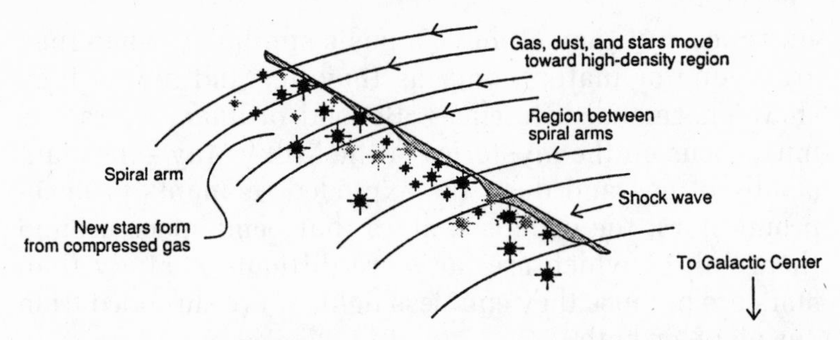


Figure 4: In a spiral galaxy, the spiral arms are the regions where young, hot stars have been born within the past few million years. A spiral density wave rotates around the galaxy's disk and compresses gas clouds to provoke the birth of these highly luminous new stars. *Drawing by Crystal Stevenson*

stars are born not at random locations but preferentially within the denser regions of a spiral pattern (see Figure 4). The pattern somehow started early in the galaxy's history and consists of alternatingly denser and less dense regions. As the pattern rotates, clouds of gas and dust that pass into the denser regions will be squeezed to some extent. The squeezing tends to trigger star formation within the clouds, and new stars appear soon (just a few hundred thousand, or a few million, years) thereafter. The result is the Andromeda galaxy—or the Milky Way, or a host of galaxies much like both these spirals.

So much for the Andromeda galaxy. After all, if we know our own galaxy well, we know Andromeda too. The difficulty is that we are inescapably (taking the medium-term view) *inside* our galaxy, unable to see it clearly as a whole, and must study it from within. Astronomers rarely say that they "can't see the forest for the trees," since they have a better metaphor at hand: They can't see the Milky Way for the stars within it. But they are far from discouraged simply because they must study the individual parts of our galaxy in order to understand what our galaxy as a whole is like.

This book deals with the parts of the Milky Way. Many of those parts are stars—about three hundred billion of them, so we shall tend to generalize. Stars come in different types, but they all have a basic similarity when they are young or mature; only as they age and die do they show enormous differences. Beyond ordinary stars, we must focus on the *mysteries* of the Milky Way—the giant clouds of gas and dust, the exploded remnants of once-proud stars, the collapsed stars that bend space around them—all of which are far more difficult to study than stars are because they emit less light, or are shrouded from our view, or both.

Thus we shall take a look at star-forming regions, at the remnants of exploded stars, at pulsars and black holes, and at cosmic beasts so strange that we can hardly say what produced them. Finally, in Chapter 13, we shall have a look at the greatest mystery of them all. Recent observations have led astronomers to conclude that most of the universe is "missing"—that is, made of matter that does not shine at all with any form of radiation. This "dark matter" apparently constitutes the bulk of the matter in the Milky Way and almost certainly the bulk of the universe as well. Astronomers have detected the dark matter solely by its gravitational force on visible forms of matter. As its name implies, the dark matter does not lend itself well to close scrutiny; we still have no good way to determine what the dark matter is, whether fish, flesh, or fowl among the suggested candidates. In short, most of the Milky Way—most of the universe—has yet to be discovered, so far as our knowledge of what it *is* goes. All we know is that the dark matter exerts the gravitational force that allows us to deduce that it *does* exist.

But let us not fear mysteries; let us rather start with the certainty that the milky way is there, even though you can't see it from the city, and that we know that this milky way consists of stars. That sentence already springs us forward through millennia of uncertainty to a firm ground of knowledge gained through technological advance and good thinking. Armed with those tools, we may proceed to examine this other Eden, the demiparadise in which the solar system circles slowly and majestically, our home galaxy, the Milky Way.

# 2

# The Gravitational Ballet

Within the milky Way—and within the entire universe—particles dance in an eternal gravitational embrace, forever reacting to the attraction exerted upon them by all the other particles in the universe. Our sun, just like its three hundred billion sisters in the Milky Way, moves in response to the combined pull of every other star in the galaxy. As it does so, the sun itself continues to exert its own gravitational tug on all its sisters—a tug whose direction changes as the sun moves. To keep track of the direction and amount of the changing gravitational forces that each individual star in the Milky Way exerts upon the solar system is a task that would baffle even our finest computers. But nature does what our computers have not yet figured out, and the combined gravitational force from all the other stars continuously "tells" our sun how to move. In short, so far as the motions of objects within our galaxy are concerned, gravity rules in the Milky Way.

## WHAT IS GRAVITY?

Gravity is a force that acts between any two objects with mass, always attracting each object toward the other. As Isaac Newton first perceived, what makes gravity so important is its *universal* character. Far from being only a local phenomenon that makes apples fall to Earth, gravity acts throughout the universe, following the fundamental rule that Newton first published in 1686.

Newton saw that the amount of gravitational force between any two objects depends on only three quantities: the mass of the first object, the mass of the second object, and the distance between their centers. Newton perceived that the strength of gravity increases in proportion to either of the objects' masses and weakens when the distance between them increases. More subtly, the force decreases in proportion to the *square* of the distance, so that if you were to move the moon twice as far from the Earth as it is now, it would feel only one-quarter as much gravitational force from the Earth. And the Earth would then feel only one-quarter as much gravitational force from the moon.

Newton's law of universal gravitation implies what may seem impossible: The moon exerts just as much gravitational force on the Earth as the Earth exerts on the moon. But note that the two forces act in opposite directions and on different objects: The Earth pulls on the moon, while the moon pulls on the Earth. Nevertheless, the two pulls of gravity are equal in strength and depend only on the product of the two objects' masses divided by the square of the distance between the centers of the objects.

How can the moon possibly pull on the Earth with the same amount of force as the Earth pulls on the moon? Still more incredible, while the Earth pulls on *you* with its gravitational force, you pull on the Earth with *exactly as much force* in the opposite direction! How can this be so? Why doesn't our planet rise to meet you halfway when you plunge to Earth from a diving board?

The answer lies in the fact that an object's mass—the amount of matter in the object—governs its resistance to acceleration as well as the gravitational force it exerts. One of Newton's greatest contributions to physics, now called Newton's second law of motion, states an utterly simple fact, just algebraic enough to appeal to mathematicians and straightforward enough to rule the universe:

*Acceleration equals force divided by mass.* Newton therefore stated that if a force acts upon an object, the object will undergo an acceleration equal to the amount of the applied force *divided* by the mass of the object.

An acceleration is a change in speed, in direction, or in both. A golfer accelerates the ball off the tee by using a club to apply a force. If he or she tees up a softball instead of a golf ball, the same amount of force will produce far less acceleration—because the softball contains far more mass than the golf ball does. What holds true for the force of a golf club holds true for gravitational forces as well. An acrobatic diver attracts the Earth—by gravity—just as much as the Earth attracts the diver. Newton showed that the diver's force on the Earth should equal the Earth's force on the diver, and Newton was right. But Newton also said that acceleration equals force *over* mass, and the diver's mass is tiny in comparison to the Earth's—about one billion-trillionth as much. Hence the Earth accelerates by almost nothing—far less, indeed, than we could hope to measure—while the diver accelerates quite noticeably, at about ten meters per second of velocity for every second the diver remains in free-fall.

Notice, by the way, that this acceleration does not depend on the diver's mass: The heavy and the light fall to Earth at the same rate. According to a lovely tradition, the great Italian scientist Galileo proved this to be so four centuries ago by dropping objects from that miscalculation of Italian architecture, the leaning tower at Pisa. Newton explained *why* this is so: An object with twice as much mass as another resists acceleration twice as much, but on the other hand, the Earth exerts twice as much gravitational force upon it. Therefore, for any object falling toward Earth, the amount of gravitational force on the object *and* the object's resistance to acceleration march in lockstep: If one is twice as large, so the other must be as well. The result is that *all* objects experience the same amount of acceleration (neglecting air resistance) when they fall toward Earth.

Newton saw that the moon obeys the same rules. If a giant hand momentarily stopped the moon in its orbit around the Earth and let it go from rest, the moon would head straight for the Earth and arrive a few days later to create a planetwide catastrophe. As the moon fell toward the Earth, the Earth would fall toward the moon, but its acceleration would be only one eighty-first of the moon's because the Earth has eighty-one times the mass of the moon, and the moon's gravitational force on the Earth equals the Earth's gravitational force on the moon. Therefore the moon would do most of the falling, but the Earth would do some falling as well.

So why doesn't the moon fall toward the Earth? Doesn't Newton tell us that gravity should produce this result? The reason why we have no such catastrophe, and expect that we never shall, lies in a key aspect of gravitation: It allows objects to fall *around* one another instead of *into* one another. Thus orbital motion under the influence of gravity plays a fundamental role in ordering the universe.

The universe is not and never has been a static place. Instead, objects in the universe are moving, and their motions affect the way the objects respond to the gravitational forces that they exert upon one another. The Earth-moon system provides a fine example. Just how the moon was formed remains a mystery, though many theories exist and conflict with one another. However, it seems clear that the moon-in-formation must have had some motion that was neither toward the Earth nor away from the Earth, but instead was in the direction *around* the Earth, that is, off to one side or another. This motion provided the key to the moon's continued existence, for it allows the moon to keep on falling around the Earth. To see just why this is so requires a look at another concept that Newton focused upon, the notion of momentum.

Momentum refers to the concept embodied in Newton's first law of motion—the fact that an object in motion tends to continue in motion in the same direction and at the same speed unless some force acts upon it. Today, when all the world drives a car, this seems clear enough: Release the car's accelerator pedal while moving at high speed and the car will coast on down the road in the same direction and at the same speed until frictional forces (or the brakes) slow it down, or a force exerted by turning the steering wheel diverts it to a different direction. In the horse-powered world of Galileo and Newton, when smooth highways were lacking, the concept seemed bolder; but these gentlemen arrived there all the same. Any object in motion has an amount of momentum measured by the product of its mass and its velocity. Velocity refers to an object's speed in a particular direction. If an object's mass does not change, any change in its speed or direction requires a change in its momentum. To produce a given change in velocity, more force must be applied to a more massive object. For this reason a semitrailer needs a stronger motor and brakes than a sports car does.

If you plan to change the moon's momentum, you need a massive object that will exert plenty of gravitational force—the Earth, for instance. Left to itself for a moment, the moon would sail off into space at whatever speed and in whatever direction that it had at that moment. Instead, the Earth's gravitational force continuously acts upon the moon to *change* the moon's momentum, basically by altering its direction of motion. In other words, the Earth's gravity deviates the moon's trajectory from a straight line into an elliptical orbit around and around the Earth.

If the moon had no momentum (an invisible hand stopping it in its tracks), this change in momentum would appear as the moon headed at increasing velocity straight for Earth. But as things are, the moon *does* have significant momentum as it moves in orbit at about one kilometer per second. The change in that momentum caused by the Earth's gravity appears as a continuous deviation of the moon's motion away from a straight line. Take away the Earth's gravity and the moon would sail directly into space. Take away the moon's momentum and it would head directly for Earth. Because both momentum and the

Earth's gravitational force exist, the moon falls *around* the Earth, month by month, eternally, as the Earth-moon system itself orbits the sun.

What is true for the Earth and moon likewise holds true for the sun and its planets. Each of the planets orbits the sun in a continuing dance of gravity and momentum, falling not into our star but around it. Each dance reflects the amount of gravitational force on the planet (this depends on the sun's mass, the planet's mass, and the distance between them) and the planet's resistance to acceleration. Like the planet's momentum as it moves, the resistance to acceleration varies in proportion to the planet's mass, so all three quantities—gravity, resistance to acceleration, and momentum—vary with the mass. It thus develops that the planet's mass does not affect its motion in orbit any more than mass affects how rapidly a diver falls to Earth. If a planet has more mass, it feels more gravitational force, but its momentum is proportionately greater and it resists acceleration in the same proportion.

The result is that only a planet's *distance* from the sun, not its mass, affects its orbital characteristics. More distant planets move more slowly in orbit than the planets closer to the sun do. The distant planets therefore take more time for each orbit, not simply because they have more distance to cover, but also because their velocities in orbit are less than those of the inner, fast-moving planets. For example, Jupiter has more than five times Earth's distance from the sun, but Jupiter takes not five but nearly twelve years to circle the sun once. Pluto, the outermost planet, averages forty times the Earth's distance from the sun as it moves along its noticeably elongated orbit, but Pluto requires 249 years to complete a single orbit.

## GRAVITATIONAL FORCES IN THE MILKY WAY

On the enormous distance scales embodied in the Milky Way galaxy, things work out much as they do in our solar system. There, too, objects—in this case stars—respond to gravitational forces from other stars, not by falling straight into anything but by falling *around* a center, in this case the center of the galaxy.

This does not imply that the center of the Milky Way contains all or even most of the mass in our galaxy. Far from it: The galactic center, though worthy of a chapter to itself, contains only a tiny fraction (about one-thousandth of 1 percent) of the mass in the Milky Way. Stars in the Milky Way orbit the galactic center not because the center has great mass but because the total mass in the galaxy *acts* as though it were concentrated at the center.

Let us be specific. Newton worked for quite a while on the calculation of how an extended object—the Earth, for example—would attract another object—the moon, for instance. Since each piece of the Earth has a different location, it might appear that Newton faced a difficult calculation. He would apparently have to calculate the force that each piece of the Earth exerts on the moon, pulling the moon with a certain amount of force in a direction that depends on just which piece of the Earth we have in mind. Newton's analysis revealed an enormous simplification: If the Earth is assumed to have spherical symmetry—that is, if the Earth does not bulge in any direction, neither at the poles nor at the equator—it acts for gravity's reign as though all its mass were concentrated in a single point, the Earth's center. In effect, the parts of the Earth to the north of the equator pull somewhat toward the north, and the parts to the south pull somewhat to the south; the total pull seems to come from the middle.

To be sure, the Earth is not quite spherical: Our planet's rotation makes it bulge a bit at the equator. As a result, Newton's simplification can be applied but does not give a perfectly exact result. The difference in effect on the moon does not amount to much, but artificial Earth satellites, which orbit much closer to Earth than the moon does, undergo changes in their orbits that result from the fact that the Earth is not a sphere. Newton's simplification can be generalized to any spherical distribution of objects. If a particle is *outside* the objects, then the objects act for gravitational purposes as if they were a single object whose mass equals the sum of the objects' masses, located at the center of the spherical distribution. But there's another wrinkle, which Newton also saw. If you are located *inside* a spherical distribution of objects, they exert *no* net gravitational force upon you. Of course, to be inside *all* the objects, you must be at the center of the distribution; you can then perceive that the pull from any given direction will be canceled by an equal pull from the opposite direction.

Suppose, though, that you are partway out from the center of the distribution of objects. You can then mentally divide the collection of objects into two parts. One part consists of all the objects *closer* to the center than you are. These objects collectively combine their pulls to exert a single force directed toward the center of the distribution and equal in strength to the force from an object at the center whose mass would equal the sum of the masses of all the interior objects. The second part of the distribution includes all the objects *farther* from the center than you are. The gravitational forces from these objects cancel one another and so exert no net force upon you (see Figure 5). So the only objects that matter—for gravitational purposes—are those closer to the center of the distribution than yourself.

## STARS IN GRAVITATIONAL EMBRACE

Consider what this means for the stars in the Milky Way. We must accept the fact that the distribution of stars does not possess spherical symmetry: Far from having a spherical shape, the Milky Way consists of a flattened disk of stars. Marvelously enough, Newton's simplification still holds true so long as the distribution of stars has "azimuthal" symmetry; that is, so long as the distribution is the

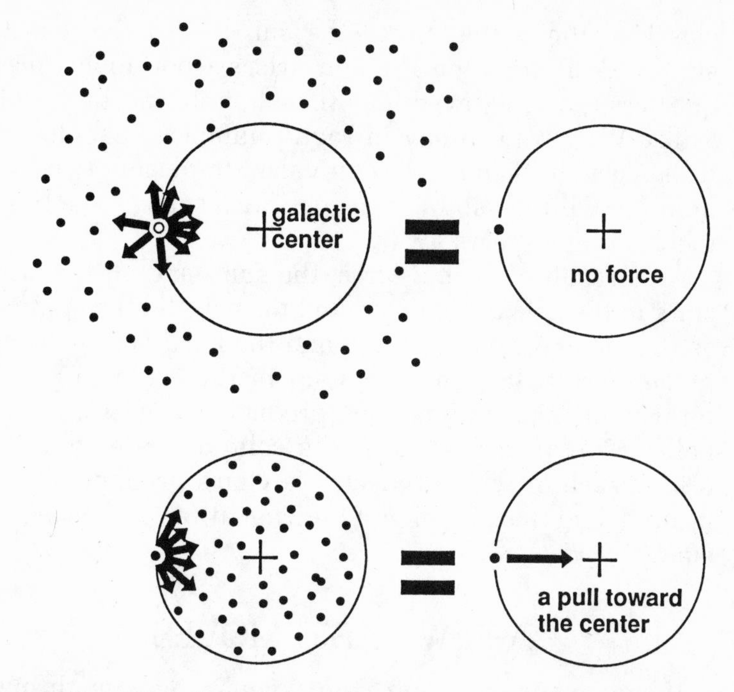


Figure 5: The gravitational pulls on the sun from all the stars farther than the sun from the center of the Milky Way cancel one another and produce a zero net force. The pulls from the stars closer to the center than the sun combine to equal the force from a single object at the center of the Milky Way whose mass equals the mass of all the stars closer to the center. *Drawing by Crystal Stevenson*

same in all directions going outward from the center along the plane of the galaxy (or in all directions at a given angle above or below the plane). If the distribution of stars has azimuthal symmetry, then stars closer to the center will still combine their forces to act like one mass at the center, and stars farther from the center will cancel one another's forces to exert no net force.

Hence a typical star—our sun, for instance—feels a net gravitational force that arises from all the stars that are closer to the galactic center and no net force at all from the stars farther from the center. The strength of this force exactly equals a force from a single object at the center whose mass equals the sum of the masses of all the objects closer to the center. Since the sun lies in the galactic suburbs, the force on the sun arises from about three-quarters (possibly even four-fifths) of all the stars in the Milky Way. We can say in round numbers that the sun feels a gravitational force equivalent to placing two hundred forty billion stars at the galactic center, some thirty thousand light-years away.

As a result of such a force, the sun—and all the other stars in the galaxy—falls around the galactic center, much as the moon orbits the Earth and the Earth and its sister planets orbit the sun. The stars in the Milky Way will forever attract one another, producing a grand cosmic ballet. This dance consists of a calm and precise set of orbits, each of which reflects Newton's principle of momentum and the mutual gravitational attraction among the stars.

## A FEW BASIC MOVES

If you want to imagine how you're moving through space, you must try to envision a triple motion: The Earth is *rotating* on its axis, *revolving* around the sun, and *orbiting* along with the sun and the other planets around the center of the Milky Way. These three basic motions occur at different speeds and take quite different amounts of time for a single cycle. The Earth rotates once each day (actually once in twenty-three hours and fifty-six minutes) at speeds that range from about a thousand miles per hour at the equator, through seven hundred miles per hour at the latitudes of the United States, to zero at the poles. The Earth orbits the sun once each year, covering about 1/365 of the distance each day. As a result, the twenty-three-hour-and-fifty-six-minute rotation period yields an average interval of twenty-four hours between successive risings of the sun, as the Earth moves partway around its orbit between sunrises (see Figure 6). The Earth's speed in orbit equals twenty miles per second, or seventy-two thousand miles per hour! But this speed falls far short of the speed

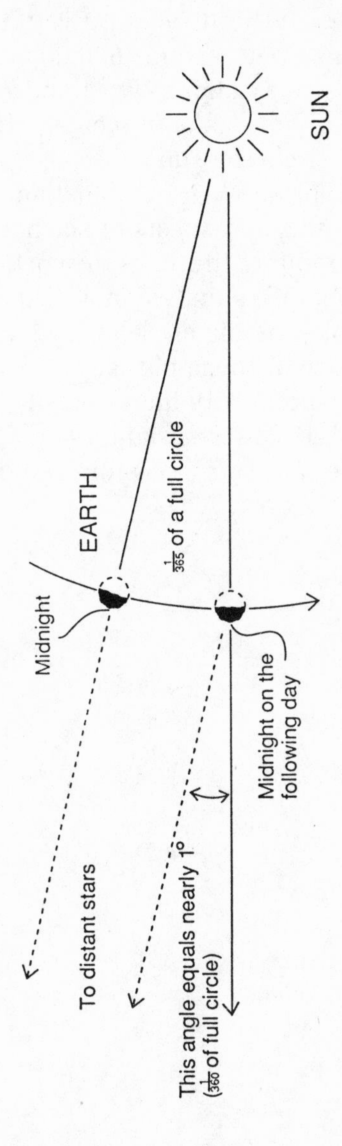


Figure 6: As the Earth rotates it also orbits the sun. As a result, it takes four minutes longer for a point on Earth to regain the same position with respect to the sun than with respect to the distant stars. *Drawing by Crystal Stevenson*

with which the solar system orbits the center of the Milky Way, thirty thousand light-years away. Proceeding at a speed of 160 miles per second (576,000 miles per hour), this grand orbit takes about two hundred forty million years, as we orbit the galactic center eight times more rapidly than we circle the sun.

Not a bad little pirouette for a planet and its inhabitants! One of the most elegant aspects of science consists of the ability to sum up many lifetimes of work in a single sentence. Try this one: We spin, we make the little circle each year, and we make the big circle every two hundred forty million years. Each of those phrases took quite a while to discover—and could hardly have been accomplished by a single individual. If you can understand what they mean, you are well ahead of our ancestors—and on the road to understanding the stars.

# 3

# THE MAKING OF STARSHINE

WE SHALL NEVER know exactly what happens inside a star. No probe could ever investigate that mystery—it would burn up or be crushed long before it entered the star. But while we cannot place a star on a table and dissect it in detail, we can observe, analyze, and model it. We can guess what is happening within the hearts of stars using what we know about nature right here on Earth. And with our best guesses we can proceed to explain what we see when we observe the stars in the Milky Way and predict what else to look for. Indeed, astronomers have demonstrated that we can extract a great deal of knowledge from very little data.

One truly overriding question has dominated human interest in stars: What makes them shine? The obvious answer is only part of the correct one. Stars shine because they are hot. But what *makes* them hot? What powers the furnaces of the stars? How does the mechanism work? And what makes it work so steadily, brightly, and virtually eternally?

There are many ways to make things hot. You can, for example, try to burn them in a conventional way—combine chemicals with oxygen and let them give off energy as they create new substances. You can also heat matter by applying friction to it. Or you can make a chemical mixture change states, as when water passes from gas to liquid, releasing heat as it does so. Perhaps you might apply tremendous pressure from gravitationally induced contraction to generate heat. You might "go atomic" and fuse light elements together or use fission to break heavy ones into lighter ones. Change a substance and it may give off energy, and if it gives off enough energy, it will generate so much heat that the substance will shine.

But try to make a fire burn for the lifetime of a typical star—ten billion years. We know of no fire on Earth that can burn close to that long. Every fire eventually goes out as it runs out of fuel. But nature has found a way to make stars bright and constant burners of the most abundant fuel in the universe.

Our observations of stars, including the sun, provide clues about what might make stars shine. If there were some way of identifying the chemicals in a star, you could see whether the star is made of gasoline, or coal, or anything else we think would fuel the star's fire. And, in fact, nature has provided a great way to identify what chemicals exist in stars through a special kind of color signature called spectral lines. These spectral lines arise because each type of atom or molecule absorbs or emits different amounts of energy from light passing through. The different amounts of energy can be recognized as different colors of the light that is added to, or removed from the light from the stellar interior. Thus the distinct pattern of colors that a particular atom or molecule absorbs or emits provides a spectral "fingerprint." In the laboratory or in a star, you need only identify the pattern of the wavelengths of these colors to know which substances are present (see Photo 4).

Why does one type of atom reveal a different set of spectral lines from another? We need to see how atoms and molecules absorb or emit light. Scientists have determined that when electrons in the atoms or molecules change their positions in relation to their respective nuclei, they do so in discrete steps. In other words, electrons can have only certain discrete orbits. When an electron jumps from one orbit to another, a "quantum" of energy called a photon, a particle of light, is exchanged. That photon carries an energy equal to the difference between the electron's first orbit around the nucleus and its next one. But different atoms and molecules have different numbers of electrons and nuclei of different weights. This changes the size of the steps that an electron in an atom or molecule can take, compared with a different atom or molecule. Different atoms or molecules have different "staircases" of steps, so in turn they have different photon energies corresponding to changes from one step to another. The result? A unique set of spectral lines for each atom and molecule.

A bit of forensics identifies a given atom or molecule. By using a spectroscope, a device that splits light or electromagnetic radiation into various wavelengths, we can look at a given substance and catalog its spectral lines. A compendium of spectral lines then allows us to perform the reverse process. By attaching the spectroscope to a telescope, we can observe the spectral lines of a star, which will tell us what's there.

We don't see the stellar spectral line fingerprints of complex hydrocarbons, which are the chemicals we usually burn here on Earth. Instead we see the spectral lines of hydrogen and helium, with weak evidence for other atoms, in the sun and stars. Hydrogen is there in the greatest abundance.

Could hydrogen be the magic fuel for stars? It's a good idea; we do occasionally use hydrogen as a fuel on Earth, most notably in some rockets. But the true burning of hydrogen, where it combines chemically with oxygen, gives off water as a by-product. It's a clean fuel—its ash a puff of water vapor. In stars, we don't see the spectral lines we'd expect from oxygen or water vapor. We know with certainty that stars do not burn hydrogen by combining it with oxygen.

Could stars generate heat through processes independent of what they are made of? A star could heat up simply because the relentless squeezing of its own gravity converts the gravitational energy into heat. Gravitational heating is a fine idea, but it turns out to be the mechanism that *induces* stars to release heat, and not the heat-releasing process itself. A simple calculation shows that if a star were to shine simply because of its own gravity, it would shrink a little each minute. If such "squeeze heating" were 100 percent efficient, the star would contract to no size at all in thousands, at most millions, of years. The existence of life on Earth over billions of years implies that the sun, and presumably other stars, does not draw its energy primarily from squeeze heating.

Another possibility on the checklist of possible ways to heat stars can be easily excluded. Perhaps the stars could change states to give off energy—for example, by turning gas in their interiors into liquid. This is similar to the process, in reverse, by which refrigerators and air conditioners work. But such changing of states is an unstable process. We must therefore explain why and how billions of stars are changing state continuously. We must also ask which changes of state would occur. Hydrogen simply can't change states in the environment of the sun, for example, because to turn hydrogen gas into a liquid requires extremely cold temperatures, hundreds of degrees below zero on the Celsius scale. That would require a very cold star indeed!

Next, consider the possibility of an atomic process such as the fission, or splitting, of atomic nuclei. On Earth, reactors based on this process produce enormous amounts of energy. Can fission stars exist? It would be difficult in the extreme to imagine a fission star made mostly of hydrogen. Fission requires that certain types of heavy elements, such as uranium, break apart into new, lighter ones. In this process, a small fraction of the initial mass doesn't become part of the lighter products. Instead, this fraction of the matter is converted directly into energy, as Einstein's equation $E = mc^2$ summarizes. ($E$ is energy, $m$ is mass, and $c$ is the speed of light.) But hydrogen simply cannot break down into lighter elements—it's already the lightest element. So if hydrogen is part of the heating process, it can't be a fission technique that's used. Perhaps, then, you could create a star that hides its heavy elements on the inside, leaving hydrogen on the outside. But such a star requires a tremendous amount of heavy elements to fuel the fission star; these unstable heavy elements typically have lifetimes much shorter than the billions of years that stars last. In other words, the fuel could be concealed from view, but it still would not be sufficiently abundant to explain the vastness of time through which stars evidently endure.

## FUSION FROM THE INSIDE

By virtually eliminating the other possibilities, we are led to the most awesome means of obtaining energy: nuclear fusion. Like fission, fusion is a process in which new elements are formed from old, with a small fraction of the original mass converted into energy. But in fusion processes, lighter parent elements are fused together to make heavier ones in a process that is almost the exact opposite of fission.

Stars work by fusing hydrogen nuclei (protons) into heavier atomic nuclei. We know from hydrogen bomb research that hydrogen nuclei will fuse to yield the next simplest element, helium, and will produce vast amounts of energy as they do so. Furthermore, in the spectra of stars we see not only lines of hydrogen but those of its fusion product, helium, as well. But these spectra also show that stars are relatively cool on their surfaces.

The sun reveals its surface temperature, about six thousand degrees Celsius, through its spectral lines and its yellow-white color. How color reveals temperature can be shown by looking at a kitchen stove. Cherry red is much cooler than white. The stove would resemble a miniature sun when it turned yellow-white. If the stove were much hotter, most of its radiation would appear in the form of ultraviolet radiation, and still hotter temperatures would make the stove a tremendous source of x-rays. At a million degrees Celsius, it would not be white but instead would emit the bulk of its radiation as x-rays. The sun's visible light results from the relative coolness of its outer layers (see Figure 7). This highlights an important fact that contradicts our intuition: The sun and all other stars are actually poor at radiating energy! Stars are places where heat is *stored* rather than *radiated*. Each star's surface layers conceal a far more radiant heart kept immensely hot by the star's own gravity.

Calculations reveal that the outer layers of the sun are nearly opaque. Stars shine because only a small fraction of the energy within them leaks out at any one time. In a sense, stars are more "dark" than "bright": Their interiors are far more energetic than the comparatively pale exteriors that we see.

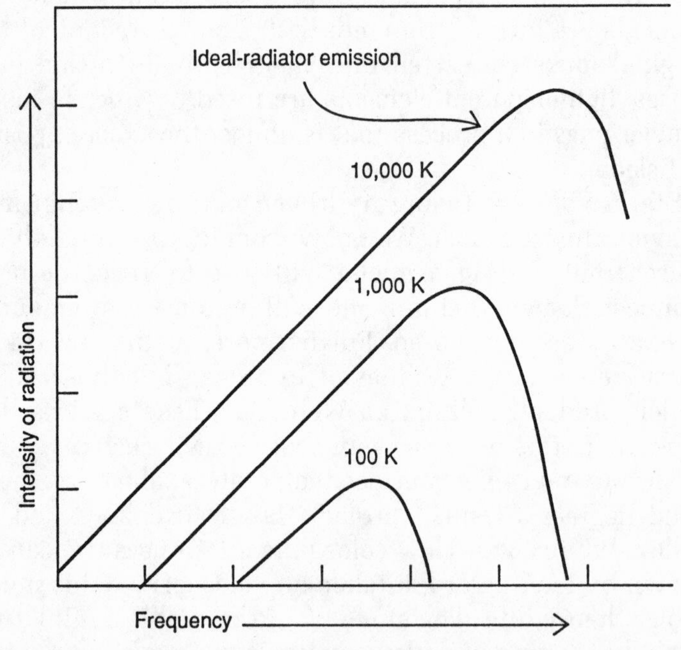


Figure 7: A black body or ideal radiator produces radiation at all energies (that is, at all frequencies and wavelengths), but the amount of radiation emitted at different frequencies depends on the temperature of the ideal radiator. Cool objects emit primarily radiation of longer wavelengths, whereas hot objects produce mostly short-wavelength radiation. *Drawing by Crystal Stevenson*

What makes hydrogen fuse in the interiors of stars? On Earth, the fusion of hydrogen nuclei is extremely difficult to achieve. First you must face the difficulty that this fusion occurs only at temperatures of many millions of degrees. It would take a million-degree match to set off a many-million-degree fire. How do stars ever reach those enormous temperatures to begin the fusion of hydrogen nuclei?

It's easy to see how stars can grow so hot at their centers once you know what happens to gas when you squeeze it: It heats up. So long as a star remains gaseous throughout, the squeezing caused by its own gravity must inevitably heat its interior, and a greater force of gravity will induce a firmer squeezing and higher temperature. Once that temperature reaches about five million degrees Celsius, hydrogen nuclei begin to fuse.

We now have all the basic pieces to the puzzle of stars. First, stars contract until their internal temperatures reach many millions of degrees, allowing hydrogen-into-helium fusion to occur. Second, the star's outer layers release only a fraction of the energy produced by the fusion furnace; they insulate the energy produced at the star's core, allowing only a small amount of that energy to escape in the form of heat. These outer layers display temperatures far cooler than those deep inside the star.

Consider a computer model for the core of a star like the sun—a set of equations solved to yield the physical conditions at each point in the star's interior. The model can be used to answer the question: How long will it take for energy released in the star's core to reach its surface layers and make the star shine?

The answer turns out to be about ten million years! Because the outer layers wrap the energy-generating core so efficiently in a nearly opaque blanket, the enormous output from nuclear fusion in the core reaches the surface not through radiation but through a tremendous number of collisions. By-products of the fusion reaction strike particles nearby; these in turn collide with particles a little farther out from the core; and through huge numbers of collisions, the energy spreads outward and heats the entire star (see Figure 8). So the light we see shining today from the outer layers of a star such as the sun arises from energy generated millions of years ago at the star's center. Indeed, if for some reason the sun's center were to cease nuclear fusion entirely, we wouldn't discover that fact in sunlight for millions of years.

Figure 8: A star is hottest at its core, where energy of mass is converted into kinetic energy. This kinetic energy diffuses outward through countless collisions among the particles in the star. As a result, the entire star grows hot, but the temperature falls steadily from the center outward to the surface as the energy spreads through a larger and larger volume. *Drawing by Crystal Stevenson*

The billiardlike collisions of countless particles represent one key way to transmit energy outward from the core. Another way relies not on collisions but on the bulk motions of hot gases within the star. The energy from the core absorbed by the layers of gas may cause this gas to seethe like water heated in a kettle. In a star the hotter layers will tend to rise above the slightly cooler ones, forming "convective cells," huge zones of circulating gas driven by the heating layers. As the hotter layers rise, they cool and sink downward, pulled by gravity. These convective cells move energy from the inside of a star to its outer regions, which radiate the energy away into the vacuum of space.

Using the combination of what they have observed, what they know about physics, and what computer models can reveal, astronomers have come to see that a star is a remarkable engine. Each star contains a fusion furnace at its heart, wrapped in hundreds of thousands of miles of opaque outer layers. The fusion engine runs on gravity, the invisible hand that squeezed the star, making it grow so hot that its center can fuse small atomic nuclei into heavier ones.

## FUSION FORMULAS

Stellar fusion can be quite versatile. For example, most stars show a correlation between surface temperature, as inferred from their spectra, and intrinsic brightness. This relationship was first found independently by two astronomers, Ejnar Hertzsprung and Henry Norris Russell; today astronomers summarize the relationship in a graph called the Hertzsprung-Russell, or H-R, diagram. The H-R diagram reveals that the more luminous stars are those with higher surface temperatures. In all stars, so long as the pressures and temperatures at the stellar core are high enough to ignite hydrogen, hydrogen fusion will occur steadily for billions of years.

Most stars simply fuse hydrogen into helium, as our sun does. But some stars have spectra that show evidence of far less hydrogen and helium than stars like our sun contain. Instead, in these stars, a variety of spectral lines from much heavier elements appear. Should we eliminate fusion as a possible method of energy production in these stars?

In truth, fusion can occur among a variety of nuclei: Hydrogen into helium need not be the only fusion possibility. Just which type of fusion will occur depends upon the conditions in the heart of a star. Those conditions can be summarized in three words: pressure, temperature, and composition. Pressure and temperature are so closely linked inside a star that temperature may be regarded as a single determining factor, along with the composition—that is, what the star is made of.

Let us consider the temperatures at which different types of nuclei will undergo nuclear fusion. When does hydrogen fuse to make helium? The main method for fusing hydrogen is a bit complicated. First, for temperatures from five million to sixteen million degrees Celsius, two hydrogen nuclei combine to form heavy hydrogen, or deuterium. Then a deuterium nucleus combines with yet another hydrogen nucleus to make light helium. Finally, two of these light helium nuclei combine to make the common type of helium nucleus and two more hydrogen nuclei. (The ash that is left actually includes a little of the initial fuel as well as the crucial new energy that arises from the slight loss of mass.) In the sun, about four hundred million tons of hydrogen are converted into helium every second. This might seem likely to exhaust the sun's supply of hydrogen, but happily the sun contains an enormous fuel supply. Since most of its mass consists of hydrogen, it will be five billion years before the sun runs out of this marvelous fuel.

Other types of fusion are activated at higher temperatures. For example, carbon nuclei will fuse, making the heavier elements of oxygen and nitrogen in the so-called CNO cycle (carbon, nitrogen, and oxygen), at temperatures of at least sixteen million degrees Celsius. Helium nuclei will fuse together to make beryllium at temperatures close to a hundred million degrees Celsius. At even higher temperatures—approaching a stupendous six hundred million degrees—carbon will fuse to make oxygen and neon. At over a billion degrees Celsius, neon and oxygen nuclei will fuse. Even heavier elements will fuse—silicon and sulfur, for example—when the temperature is hot enough, say a billion and a half degrees (see Table 1).

If these formulas exist, why don't they all become activated simultaneously inside the sun or any other star? Temperature is the most important factor, because a star is

**STAGES OF NUCLEAR FUSION IN A MASSIVE STAR**

| Name of Stage | Fuel | Final Ash | Temperature required (Millions of K) |
|---|---|---|---|
| Hydrogen burning | Hydrogen | Helium | 5 |
| Hydrogen burning | Carbon, Nitrogen Oxygen, Hydrogen | Helium | 16 |
| Helium bruning | Helium, Beryllium | Carbon | 100 |
| Helium burning | Helium, Carbon | Oxygen | 100 |
| Carbon burning | Carbon, Neon, Magnesium | Oxygen | 600 |
| Neon burning | Neon, Helium | Magnesium | 1,000 |
| Oxygen burning | Oxygen | Sulfur, Silicon | 1,500 |
| Silicon burning | Silicon | Iron | 3,000 |

Table 1: Different types of particles can fuse together to produce kinetic energy from energy of mass. Each type of fusion reaction requires a different temperature to proceed.

the ultimate self-regulating thermostat, continuously adjusting its interior in response to changing conditions there. The constant squeeze of gravity keeps the star's internal temperature high. But fusion keeps the temperature from getting *too* high. Simply put, the energy liberated by the fusion itself exerts an expanding pressure. This counteracts the squeeze of gravity, preventing the core pressure from rising so high that the temperature grows much larger than that needed to make hydrogen fuse.

That radiation pressure makes a star literally buoyant against its own weight. As soon as the radiation pressure becomes strong enough, it will slow, and then stop, a star's contraction. The radiation will push out against gravity's inward pull. So long as the star is consuming its abundant supply of hydrogen, that buoyancy will be maintained and the temperatures at the fusion furnace will remain a mere million degrees Celsius.

Yet different temperatures from the star's mass will provide for variations of the hydrogen fusion formula. For example, astronomers' calculations have revealed and confirmed the fact that a minimum of about seven one-hundredths of the sun's mass is needed to make nuclear fusion begin. With less than this mass, the star will never produce temperatures hot enough to light the hydrogen fusion furnace. At the other extreme, however, hydrogen will still burn well in stars with two hundred times the sun's mass. Above that mass, the buoyancy becomes unstable and the controlled explosion of fusion risks becoming an enormous hydrogen bomb.

Based on determinations of stellar masses made by observing stellar motions, astronomers conclude that most stars visible to us fit nicely into this range of masses. But if all stars have the proper masses to fuse hydrogen nuclei together, when might a star *not* burn hydrogen?

The answer lies in the exhaustion of hydrogen fuel at the furnace. Consider the accruing ash—in this case, the by-products of hydrogen fusion: helium, deuterium, and so on. Each such ash will form a core in the hydrogen fusion furnace. Through most of a star's lifetime, the only active role in the energy of the star of these regions of ash will be to help move energy away from the core. They will not, themselves, be part of the burning. But in one critical case the situation changes—when the hydrogen furnace falters as it runs out of fuel (hydrogen nuclei).

When the hydrogen furnace grows exhausted, it loses some of its hard-won radiation pressure. As a result, the star contracts. The contraction produces higher temperatures and pressures, and this causes the small remaining amount of hydrogen fuel to burn more swiftly. Once the contraction produces temperatures approaching one hundred million degrees Celsius, the former ash ignites: Helium nuclei begin to fuse together. At this stage, the central fusion furnace is a helium one, running at a much higher temperature than before, and the star becomes its own breeder reactor, consuming its own ash as new fuel.

All stars are composed mainly of hydrogen, even helium-burning ones. The hydrogen has not disappeared in these stars; it has simply been exhausted at the star's center, in the furnace itself. The star's outer layers, still nonparticipants in the fusion process, will remain mostly hydrogen, while the core runs on burning ash.

A helium-burning star, by its nature, must be quite old. After all, almost all the hydrogen at the core must be used up before the helium takes its place as fuel. Computer simulations show that up to several billion years may pass before an ordinary star forsakes its hydrogen-burning core for helium.

But the helium furnace shares the core with a new *hydrogen* furnace that surrounds it. By the time the contracted star's hotter temperatures have ignited the helium fusion, the regions immediately surrounding this core still include plenty of hydrogen, which until now was not hot enough to fuse. As the helium core contracts, however, the shell of hydrogen-rich material surrounding the core ignites as a *second* fusion furnace. The helium-burning star now has two fusion furnaces, like a stellar layer cake of concentric fusion engines.

In a star with ten or twenty times the sun's mass, the cake develops more layers. Each layer represents a fusion furnace with a new fuel supply. As each type of fuel grows exhausted, the star contracts to produce higher and higher temperatures at its center. Eventually the ash product of the inner furnace will ignite; as a result, the star resembles an onion that records bygone fusion processes. The heaviest nuclei—iron, for example—appear at the center, while outside lie layers of progressively lighter nuclei, such as silicon, magnesium, neon, oxygen, and helium. By the time the core produces its heaviest ash, iron, it will have finally made an ash that will *not* burn. Iron nuclei simply yield no energy when the fuse together; instead, this fusion *consumes* energy. By the time a massive star's core has become mostly iron, its temperature is well over a billion degrees Celsius. From there it declines to a few million degrees for the cooler outermost hydrogen-burning layer. The hottest, brightest fusion furnaces turn out to have the shortest lives. The final stage, the fusion of silicon nuclei to form iron, takes only a few *weeks*. Eventually the star's outer shells must outlive the hotter furnaces inside. We shall see the result in Chapter 12.

## FROM THE COOLEST, THE HOTTEST

What evidence exists that all these fusion processes occur in stars as we have described them? The evidence is indirect but intriguing. It consists basically of astronomers' ability to relate the different types of stars they see to their calculations of what happens to stars as they age. We have already discussed hydrogen-burning stars. Curiously, the models that predict the shells of fusion in an older star also predict that the star will acquire a strange appearance. As the star's core contracts, its exterior will cool and expand. The star's hydrogen-burning shell will push its outer layers outward to vast distances, perhaps well over five hundred million kilometers from the innermost core. The layers will cool because they are so far from the energy-producing center, and they present so large a surface area that all the energy passing out from the core furnaces will be radiated away at low temperatures, perhaps only one-third the surface temperature of our sun.

We can easily see such bloated stars. We call them red giants and red supergiants for their red color and huge sizes. The spectra of the light from red giants reveal the presence of other elements in addition to hydrogen and helium. The most abundant of these are carbon, oxygen, and nitrogen. Red supergiants have enough mass eventually to ignite all the furnaces; red giants have just enough mass to ignite helium and hydrogen.

Betelgeuse, the second-brightest star in the constellation Orion, is a red supergiant, believed to be burning helium at its core. Its size is legendary; if our sun were as large, its outer layers would stretch beyond the radius of Mars's orbit. In other words, Earth would be *inside* our sun if our sun were such a red supergiant. Betelgeuse is one of the few stars beyond the sun whose appearance is anything other than a point of light; thanks to its enormous size, we can actually photograph the disk of its surface layers (see Photo 5). Betelgeuse shows changing regions of relative cold and hot. But even at its warmest, the star's surface layers never reach three thousand degrees Celsius, half as hot as the sun's surface, and belie the one-hundred-million-degree helium furnace within.

We see red supergiants that are even more massive—scaled-up versions of stars like Betelgeuse. But red supergiants last only a few million years; their inner furnaces can be unstable and may expose the hydrogen to superhot temperatures. Red supergiants such as Betelgeuse seem to be losing a tremendous amount of gas and dust to space. Some stars (the less massive ones) push out their outer layers and can even lose them. We see the result as planetary nebulae, beautiful, expanding shells of hydrogen and other gases (see Photo 6).

But still more massive supergiants face a more violent fate. Once their billion-degree cores become mostly iron, the stars cannot hold up against gravity. They collapse and explode violently, as discussed in Chapter 12.

And what will happen to the sun? For a few billion years, not much will change. But eventually, the sun will puff out its outer layers, or shed them in planetary nebulae. By then, life on Earth will long since have ended, since the sun's red-giant phase will evaporate the Earth's oceans and leave our planet as a cold piece of ash orbiting a once-brilliant star.

# 4
# The Great Molecule Factory

Like human beings, stars are born, grow old, and die in a continual process of creation and dissolution. From an astronomical point of view, the chief fact about the visible universe is that a multitude of luminous gaseous spheres shine for millions or billions of years, then either fade into obscurity or die in fiery explosions. If you want to understand the universe, you must start by understanding stars and their life cycles, and if you want to understand stars, stellar nurseries provide a good place to begin. Such nurseries, each of which typically spawns hundreds of thousands or even millions of stars, appear throughout the disk of the Milky Way galaxy and probably number more than a thousand at any given time. Of all these stellar breeding grounds, the nearest and dearest to astronomers is the Orion Nebula (see Photo 7).

## ORION: BIRTHPLACE OF STARS

On a clear night in fall and winter, you can easily look across a few thousand trillion miles of space to watch newborn stars light their cocoons of enveloping gas. To do this, you must locate one of the largest and brightest constellations, Orion the Hunter, which spreads over more than twenty degrees—as large as the Big Dipper. Ancient Greek mythology tells of the Hunter who, followed by two hunting dogs, was placed in the sky as a constellation once he had been bitten to death by Scorpius the Scorpion, which is also a large constellation but located at nearly the opposite side of the sky. When Orion is above the horizon, for any observer anywhere on Earth, Scorpius is below, and once Scorpius has risen, Orion has set. Hence the Hunter need never see his nemesis, nor can the Scorpion gloat over his victim.

Orion is to casual star watchers the most satisfying of all the constellations because it is nearly the easiest to identify, and it contains as interesting a group of objects as any region in the sky. To identify Orion, look for his bright belt of three rather closely spaced stars. These three stars—Alnilam, Alnitak, and Mintaka, to use the Arabic names given to them a millennium ago—rank among the most intrinsically luminous, most prodigious producers of energy that you can easily see in the Milky Way. To the north of Orion's belt lies Betelgeuse, the Hunter's brighter shoulder, a red supergiant about five hundred light-years away. Orion's fainter shoulder is Bellatrix. To the south of Orion's belt lies the Hunter's bright foot, the blue supergiant star Rigel, even brighter than Betelgeuse, a young, hot star more than eight hundred light-years away, shining with light that left around the time that Saladin grappled with the crusaders (see Photo 8).

The three stars in the belt, the two bright shoulders, and the one bright foot (Rigel) are the most easily recognized parts of Orion. In addition, Orion's fainter foot, Saiph, can be seen on a clear night, as can a host of stars near Orion's shoulders. Most of these are taken to represent Orion's cloak, which the Hunter is flinging over his mighty pectorals as he travels the sky.

Orion is followed celestially by Sirius, the brightest star in the sky and heart of the Big Dog, Canis Major; and by Procyon, a star nearly as bright and one of the two stars visible in the Little Dog, Canis Minor. But once you have identified the main contours of Orion, you can appreciate an additional feature—the sword that "hangs" below the belt, reaching toward the Hunter's feet. This sword consists of three stars, of which the middle one is not a star at all, but the Orion Nebula, an enormous stellar nursery whose inner heart we have only begun to observe, and which will soon—in a few million years—give birth to a star cluster.

## THE ORION NEBULA

The Orion Nebula has been recognized as an unusual object since the late eighteenth century, when Charles Messier, an amateur astronomer dedicated to the search for new comets, decided to list fuzzy-appearing objects that might otherwise deceive him into thinking that he had discovered a comet. Today Messier's fame resides in his list of about a hundred rejects, all noncomets, which have turned out to include gas clouds and star clusters within our own Milky Way galaxy and a host of galaxies far beyond our own. Much of the history of nineteenth- and early twentieth-century astronomy consists of attempts to understand the nature of the objects on Messier's list—Milky Way mysteries of bygone days. Older astronomers still often refer to celestial objects by their Messier numbers; the Andromeda galaxy, for example, trips from their tongues as M31 and the great globular star cluster in the constellation Hercules is, to them, M13.

The Orion Nebula is M42, sixteen hundred light-years away, shining with light from stars that are less than a million years old. But you rarely hear astronomers call this region M42 because a certain confusion has set in—a healthy confusion that results from recent discoveries about this star-forming region. Astronomers now know that the *visible* Orion Nebula—the object that Messier numbered 42—forms only a small fraction of the material in the star-forming region. When an astronomer says "Orion Nebula," he or she is speaking about something much larger and more important than the object Messier saw, or the object visible in an ordinary telescope (see Photo 9). The Orion Nebula, with its delicate wisps of

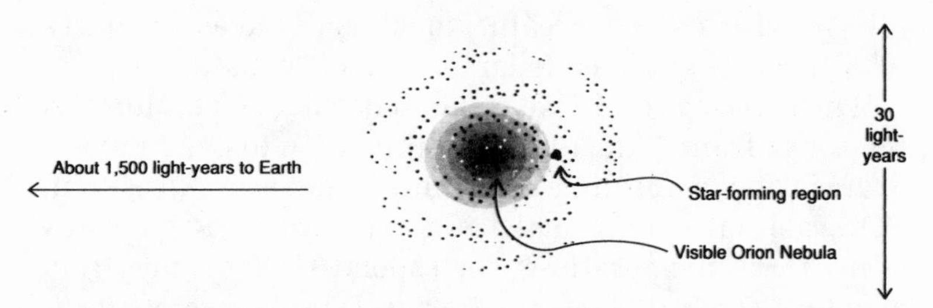


Figure 9: Around the visible Orion Nebula, a much larger cool cloud of gas and dust containing more than a million times the amount of matter in the sun may eventually form into stars. *Drawing by Crystal Stevenson*

glowing gas and warm dust, is merely the visible-light tip, enormous enough as it is, of a much larger gaseous iceberg, the cool, dense, and dark caves of gas in which stars are forming (see Figure 9).

## HOW COOL IS COOL? HOW DENSE IS DENSE?

Of course, everything in the universe is relative. When we say "enormous" for the size of the visible Orion Nebula, we mean that the nebula spans a few dozen light-years—about one-thousandth of the distance from the solar system to the center of the Milky Way. When we say "much larger" for the size of the dark matter that surrounds and enfolds the Orion Nebula, we mean one or two hundred light-years across, not even 1 percent of the distance to the galactic center. When we say "dense" for the material in this enormous, dark region, we mean to imply a density anywhere from a thousand to a million times greater than the average density of gas in interstellar space. Large though these factors may be, they still leave the gas far less dense than air. To be more precise, a density one million times greater than the average density of interstellar matter provides us with matter only one hundred-trillionth as dense as air. You could fill a battleship with this "dense" gas and still not have as much mass as you take in with a single breath. Nonetheless, to astronomers who specialize in interstellar gas, this *is* dense.

When we say "cool," we mean to denote temperatures in the range from 50 to 100 K—that is, fifty to one hundred degrees above absolute zero. Since degrees on the Kelvin (K) scale have the same size as degrees on the Celsius scale, these temperatures correspond to the range from about −223 to −173 degrees Celsius (absolute zero falls at −273.15 degrees Celsius), hence to temperatures of about −400 to −300 degrees Fahrenheit. By Earthbound standards this is not merely cool but decidedly cold; however, since much of the universe has temperatures noticeably closer to absolute zero, astronomers call these temperatures "cool" to denote the fact that *something* has kept them at their relatively elevated levels rather than allowing the regions to cool down to 10 or 20 K, or even lower.

## THE MYSTERY OF STAR FORMATION

No one doubts that stars do form: The proof lies all around us, and nowhere more strikingly than in the Orion Nebula. There approximately one thousand stars have been born within the past million years. We know the time limit because some of these young stars are so luminous that they can *last* no longer than a million years in all. But how did these stars begin to form in the first place? What makes a cloud of interstellar gas and dust give birth to a few thousand, or a few million, stars?

Nobody knows; the secret of starbirth remains hidden in the shrouds of stellar nurseries. Astronomers know that the secret has an intimate connection with gravity, but the details remain an enigma. If a clump of gas becomes significantly denser than its surroundings, then that clump's own gravity—the attraction of each piece of matter in the clump for all the other pieces—will make the clump pull itself together: The clump will contract through self-gravitation because its own gravity dominates the situation. Since the clump has a higher density of matter than its surroundings, the clump's own gravity defeats any tendency of the surrounding matter to maintain the clump against contraction. Hence the clump grows smaller and denser until its central regions become so hot and dense that nuclear fusion begins and a star is born.

That is the easy part—once you have the clump. But what processes make clumps denser than average within an interstellar cloud? And what processes made the interstellar clouds themselves—clumps on a larger scale? An interesting speculation assigns to exploding stars (supernovae) the responsibility for ruffling an interstellar cloud, making some of the cloud denser than average and thus triggering a burst of star formation within the cloud. This theory has a certain appeal—but since exploding stars were themselves born within interstellar clouds, there is a first-cause problem: What made the clumps that eventually produced the stars that exploded before any exploding stars existed?

We have no good answer. Clearly, if we hope to learn more about star formation, the most direct way to do so would be to observe the process directly. But here there is another problem: The same dense gas that turns into stars blocks all visible light, and we have no hope of observing star formation with an ordinary telescope. Better opportunities await us with longer-wavelength radiation—infrared, microwave, and radio. These rays can penetrate through gaseous, dusty regions much better than visible light can. Of these three types, probably the one with the most potential—certainly the one with the most *unrealized* potential—is infrared.

## INFRARED PHOTONS TO PIERCE THE DARK

Most of the gas in interstellar space is cool, dense, and dark. We have taken the trouble to define "cool" and "dense" and must now determine just what we mean when we say "dark." Presumably we mean—just as one would suspect—that dark means *dark when we observe in visible light*. But visible light forms just one portion, and a tiny portion at that, of the total spectrum of electromagnetic radiation. An object "dark" in visible light may be emitting other types of radiation that we cannot observe without help—without the detectors that astronomers have spent impressive amounts of time, energy, and money in developing, improving, using, redesigning, and reimproving. Of all types of detectors for electromagnetic radiation, infrared detectors rank among the most difficult to create. Infrared photons each have relatively little energy, and therefore cannot do the spectacular things to atoms and molecules that more energetic gamma rays, x-rays, ultraviolet, and even visible light can—things that allow us to detect these higher-energy photons more readily by the effects they produce. On the other hand, unlike radio photons, which readily make faint electrical currents—currents that we now know how to amplify and to study with incredible precision—infrared photons don't do much to produce a current. Speaking with greater scientific accuracy, infrared photons *do* have a noticeable effect when they strike certain crystals. The problem is that *everything around us is radiating infrared photons like crazy*. This makes for trouble when you want to detect infrared radiation from a faraway source and not from your surroundings.

Why does everything radiate infrared? The reason is that "everything"—our neighborhood here on Earth—has temperatures of a few hundred K (degrees above absolute zero), not just a few K or a few thousand K. Nature has arranged that every object not at a temperature of absolute zero radiates photons at all wavelengths and frequencies. As a result, any object steadily loses energy through the electromagnetic radiation that it produces. If you wait for a while, say for a few thousand or a few million years, any object not resupplied with energy will cool down, eventually to temperatures close to absolute zero. The entire universe faces such a "heat death" if it continues to expand forever, since each object within it will slowly but steadily radiate its energy into an ever-increasing volume of space.

At a given temperature, any object radiates *primarily* within a certain band of wavelengths and frequencies. Technically, the object produces some photons at every wavelength, but as a practical matter, most of the photons are radiated within only one or two of the broad regions that we have seen to characterize the spectrum. Objects at temperatures of millions of K radiate primarily gamma rays; those at hundreds of thousands of K radiate x-rays; at tens of thousands of K the radiation is primarily ultraviolet; in the low thousands of K visible light predominates; for objects at hundreds or tens of K infrared radiation is the chief output; and at temperatures below 10 K, radio, or at least its shorter-wavelength component, microwaves, gets the action.

Room temperature, approximately 20 degrees Celsius, corresponds to 293 K. The hottest temperature recorded on the Earth's surface, about 50 degrees Celsius, equals 323 K, and the coldest, about −70 degrees Celsius, brings us no lower than 203 K. In short, Earth has a temperature of a few hundred K, within the range at which objects radiate primarily in the infrared part of the spectrum. Had we infrared-detecting eyes, daylight and nighttime would make little difference to us, for the entire landscape—trees, rocks, birds, buildings, and people—would emit an infrared glow. On another planet, an ever-vigilant providence or an ever-adapting evolutionary sequence may well have produced infrared eyes. On our own planet, the closest evolutionary product consists of certain snakes, known as pit vipers, whose heads incorporate special infrared-sensing regions that help them to locate their prey. We do, of course, each have a large infrared-sensitive organ: our skin. However, this did not help much when you—or your prehistoric ancestor—sought a meal or aimed to avoid becoming one: By the time that you noticed another infrared-emitting object by means of the warmth it produced on your skin, the scene would have changed, with potentially fatal results.

What nature has left undone on Earth, astronomy must do for itself. Hence astronomers have developed infrared-sensitive devices. In this quest they are not alone: Military organizations around the world have spent lavishly to develop infrared eyes for their soldiers and sailors; they know that so long as every object emits infrared, one of the best ways to be the hunter and not the hunted consists of observing the other fellow's infrared emission before he observes yours. Infrared sensors have come a long way during the last two decades, and they will doubtless improve a good deal more before the century ends. One benefit of this development has been the creation of a new class of astronomers—the infrared astronomers.

## INFRARED ASTRONOMY

Infrared astronomers tend to be instrument-oriented observers of the cosmos who don't mind spending a few weeks or months improving an infrared detector that may give them a better look at what has never been seen before—the radiant sky in infrared. But the growth of infrared astronomy, and the employment of infrared astronomers, has been hampered by a serious problem: The Earth's atmosphere blocks all but the shortest-wavelength infrared radiation. The military can do fine with this small segment of the total infrared spectrum, since even that small portion with the shortest wavelengths will suffice to reveal the enemy (or the friend, for that matter). But astronomers want to observe the universe as completely as possible, and are certainly not content with only these few percent of the shortest wavelengths in the infrared spectrum. This discontent sharpens once you realize that the shortest-wavelength infrared contains relatively few spectral features—the absorption and emission at certain definite wavelengths by particular types of molecules that provide a cosmic "fingerprint," enabling astronomers to deduce the composition and temperature of faraway clouds of gas and dust.

Because of the absorption of infrared radiation by the Earth's atmosphere, a great dream of infrared astronomers has been the creation of an infrared-observing satellite capable of observing the universe in the radiation that all cool objects emit. That dream came true for ten glorious months in 1983, when the *Infrared Astronomy Satellite* (*IRAS*), a creation of the United States, the United Kingdom, and the Netherlands, used its infrared detectors to make the first true surveys of the sky in infrared (see Photo D). *IRAS* also made narrow-angle, higher-magnification observations of regions of special interest; one of its finest products was an infrared image of Orion (see Photo C). This image clearly reveals that essentially all of the Orion region contains cool interstellar gas and therefore has the potential to form stars someday. The southern half of Orion—the region centered on the Orion Nebula—shows especially strong infrared emission, the result of a particularly dense concentration of interstellar gas.

*IRAS* was a fine satellite, but we must use the past tense, because *IRAS* no longer works. Any infrared detector has a big problem: its *own* infrared emission. Since all "cool" objects produce most of their radiation in the infrared, an object at room temperature or even at −100 degrees Celsius emits so much infrared radiation that the detector tends to be saturated by its own emission, like a heavy smoker trying to study a distant signal fire. The only good solution to this problem is to reduce the amount of infrared emission of the detector and its peripheral equipment, and the only way to do that is to *cool* the detector as much as possible. *IRAS* had a supply of liquid helium, cooled to 4 K (−269 degrees Celsius), that flowed around as much of the instrumentation as could be arranged by *IRAS*'s architects. The helium did a fine job of cooling, allowing *IRAS* to make its excellent observations, but because the helium had to be allowed to circulate over the detectors, it slowly evaporated. The evaporation robbed *IRAS* of the chance to stay cool, and by November 1983 the party was over. *IRAS* now circles the globe with a noble past but no present infrared-observing capability.

Undaunted, infrared astronomers in the United States have a new dream: the *Space Infrared Telescope Facility* (*SIRTF*). *SIRTF*, which may orbit as early as 1999, will have a larger mirror, a more sensitive detector system, and a better cooling apparatus than *IRAS* had. Properly maintained, *SIRTF* should last for years, and astronauts can ride the Space Shuttle from time to time to replenish the evaporating helium. *SIRTF* should eventually answer the prayers of infrared observers by providing an ongoing means to study the cosmos in most of its infrared emission. Meanwhile, astronomers have the results of *IRAS*'s infrared observations, plus a host of additional detail.

## INTERSTELLAR MOLECULES IN ORION

Detailed knowledge about the chemical composition and the motions of the interstellar gas in Orion can be obtained from observations of particular types of molecules in the star-forming regions. Molecules each consist of two or more atoms held together by their electromagnetic forces. Each type of molecule, like each type of atom, can absorb photons with certain definite wavelengths and frequencies. This means that if radiation from a source behind an interstellar cloud passes through that cloud on its way to us, astronomers can study the pattern of absorption lines that the molecules in the cloud will produce in the radiation. The pattern allows them to determine which types of molecules and how many of each type are contained in the interstellar cloud. But what if the astronomers don't *have* a "more distant source"—an object behind the cloud that conveniently emits streams of radiation that pass through it? Then they must hope that the cloud will *emit* radiation on its own, rather than *absorb* radiation from behind.

Otherwise astronomers have no way to determine what types of molecules exist in the interstellar cloud.

A star fulfills the requirement of producing energy by itself, and astronomers have no trouble studying the chemical composition of stars by examining stars' visible-light spectra. But a cloud of gas and dust—a star-forming region, for instance—tends to emit almost no visible light. Instead the cloud produces primarily infrared radiation, and we have seen that our atmosphere blocks most of the infrared from view. But there's still hope: The molecules within the cloud emit photons at many radio and microwave frequencies as well as infrared. The radio and microwave photons pass out through the cloud rather easily, and they pass through the Earth's atmosphere too. If you have a radio telescope, therefore, you can observe radiation at particular wavelengths and frequencies characteristic of the particular types of molecules within interstellar clouds. You will do especially well if you have a *microwave* telescope and detector because most of the molecules emit primarily at microwave wavelengths rather than at the somewhat longer radio wavelengths. (Atoms, because they are less complex than molecules, typically do not emit at radio or microwave frequencies and wavelengths.)

Using this general principle, radio astronomers have observed dozens of different types of molecules—water, carbon dioxide, silicon carbide, carbon monoxide, and sulfur dioxide among them—in star-forming regions such as the Orion Nebula. By now the number of different types of molecules known to exist in interstellar clouds exceeds six dozen. Intriguingly, astronomers are now beginning to find molecules typical of the basic building blocks of life itself. For example, long-chain molecules that use carbon as the key connecting link appear both in interstellar space and in all forms of life. Astronomers who once dismissed as pure imagination Fred Hoyle's science-fiction novel *The Black Cloud*, in which an interstellar cloud turns out to be alive, are now willing to discuss seriously the question of whether life could originate in interstellar clouds. We may yet find—though this must still be called a long shot—that complex molecules in the Orion Nebula have assembled themselves to the point that primitive forms of life (to speculate mildly) exist there.

Meanwhile, let us review the situation that surrounds the Orion Nebula. The visible nebula forms only a small part of what astronomers now call the Giant Molecular Cloud (GMC) in Orion. *Giant* means what it implies, and *molecular* means that most of the material in this region has formed molecules rather than the unconnected atoms that dominate the less dense regions of interstellar space. A star-forming region such as the one in Orion is a giant chemical factory, joining individual atoms into molecules in huge numbers. Indeed, the chemical reactions within this stellar nursery are so exotic (by our standards) that Orion makes types of molecules we ourselves can't produce in our laboratories when we attempt to simulate the conditions that we think exist in interstellar space.

The Giant Molecular Cloud in Orion contains millions of solar masses of material—the raw stuff for a giant star cluster. Within this GMC a few *thousand* solar masses—at most a few tens of thousands—have already formed stars. These stars light their stellar nursery from within, creating the Orion Nebula that Messier first numbered. But the Orion Nebula includes much less than 1 percent of all the material that might form stars in that part of the Milky Way.

For now, we don't know whether the bulk of the Orion GMC will remain relatively diffuse and never form stars, or whether it too will soon (that is, in a few million or tens of millions of years) turn into stars, creating not a small but an enormous star cluster. We do know that many astronomers will devote themselves to a better understanding of both the composition and the dynamics of the material in Orion, hoping to resolve the fundamental mystery of stellar birth: What makes stars form?

# 5
# Star Pox

If the sun were moved away from Earth to the distance of the nearest stars, it wouldn't look like much. Dim and inconspicuous, it would appear to be just another twinkling pinpoint of light, shrinking from its present angular size of half a degree to a speck a million times smaller.

On that distant scale, the seething fury and the details now visible on our violent sun would disappear into that tiny speck of light, reminding us that the sun offers our sole opportunity to study a star close up. Other stars cannot provide as much information to astronomers simply because they are so distant. Even with tomorrow's technology we shall never see more than a handful of giant stars as anything but points of light. We must continue to rely on the detailed view of a star that our sun gives us every day. Even so, some stars provide us with hints of spectacular, enormously more violent, examples of what goes on in the sun. We can therefore appreciate some of the sun's more capricious behavior as a pale example of what happens in some special classes of stars.

## SOLAR VIOLENCE

On a scale visible to our eyes, the sun appears relatively unchanging. But in astronomical terms, only a moment's wait will show change. A solar telescope with appropriate filters and a coronagraphic disk, which produces an artificial eclipse, reveals the dance of the sun's roiling, boiling atmosphere over minutes, hours, and days, as loops and delicate arcs break away from the edge of the seething sun, jetting outward for tens of thousands of miles. Many of these prominences dwarf the Earth in size. Most fall back only to become part of yet another eruption. Occasionally a giant arc shoots out with so much momentum that it escapes the sun's gravitational grasp. It forms a pocket of superhot hydrogen gas that slowly cools as it escapes into space, like the far-flung magma from a volcanic eruption (see Photo 10).

From the immediate surroundings of the solar surface flows a stream of particles called the solar wind. The "wind" consists of protons, electrons, and ions (atoms stripped of one or more electrons), all of which speed outward at several thousand miles per hour to regions far beyond the Earth's orbit before they merge with other particles in interplanetary space. Earth's atmosphere shields us from this onslaught in a colorful way, and its indirect effects are remarkable, especially when great "flares" near the solar surface turn the usual wind into a real "hurricane."

Streaming in at thousands of miles per hour, the particles in the solar wind encounter the Earth's magnetic field, which guides the particles toward the two magnetic poles. The charged particles collide with molecules high in the atmosphere, breaking some apart and ionizing some of the atoms. The energy from the solar-wind particles makes atoms and molecules glow in soft colors—greens, reds, and grays—that produce the aurorae, better known as the northern and southern lights. Thanks to the guiding effect of the Earth's magnetic field, the aurorae are most commonly seen at locations with high latitudes.

The auroral glow arises some fifty miles above the Earth's surface. At lower altitudes, the solar-wind particles can produce a geomagnetic "storm," disrupting all shortwave radio communication and producing pulses of low-frequency radio energy, which can all too easily be picked up by power lines. This radio energy can create a surge at power stations that power transformers can't handle. Thus even the effects of a distant solar storm can disrupt our everyday lives.

We now know enough about the sun to predict, on a decade-by-decade basis, when such events are most likely to happen and when the "quiet" sun will temporarily become a most "active" sun, when even the giant loops and arcs in solar prominences are dwarfed by the huge fingers of solar flares, far more energetic—and "windy"—than the loops and arcs.

Astronomers refer to the changing patterns in the sun's overall behavior as the sunspot cycle. The cycle indicates the number of dark spots that mar the brilliant face of the sun over time. Centuries of observation have revealed that the number of such sunspots follows an eleven-year-long pattern (see Figure 10). At the peak of each sunspot cycle, the sun is far from sedate; loops and arcs occur more often and are typically larger and more energetic than at other times. From large groups of sunspots, flares—tendrils of hot gas, sometimes in streams as long as the distance between the Earth and the moon or more—often erupt. Because sunspots are correlated with giant flares erupting from the sun's surface, the sunspot cycle also reflects the cycle of intense solar activity that can produce auroral displays and threaten our radio communication.

From the standpoint of the entire population of stars, however, the sun can only be called sedate. Fossils from

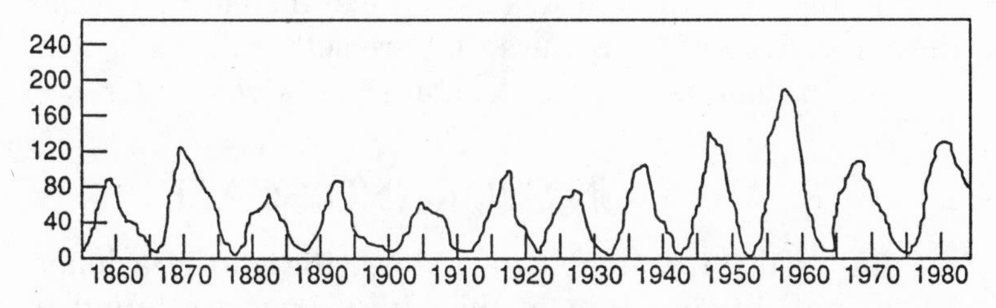


Figure 10: The average number of sunspots changes repetitively over a twenty-two-year cycle, with a maximum and a change of magnetic polarity every eleven years. *Drawing by Crystal Stevenson*

the past imply that the sun has been shining at a nearly constant rate during at least the past half-billion years, almost unaffected by the sunspot cycle or other changes. Not every star must behave in this manner, and not all stars do so.

For life that evolved beneath a sedate star, as life on Earth has, a more violent sun would be a stellar Jekyll and Hyde that would create hell on Earth. Rather than merely causing periodic geomagnetic storms or bad radio conditions, a more violent sun would periodically rip part of the Earth's atmosphere away and would tend to destroy all organic compounds—all life—in the process. Solar flares from a violent sun could easily reach the Earth, melting its surface and vaporizing its soil. A stronger solar wind could slowly change the Earth's orbit, which would over time become more elliptical, causing unbearable extremes of hot and cold each year. Eventually the Earth's orbit could become so elongated that tidal forces from the sun would break the Earth into pieces that would spiral into the sun itself. Only for a second would the sun register the change caused by the infall of these planetary tidbits.

We may pause to salute our constant sun, changing only modestly throughout its sunspot cycle. Most stars probably have only modest sunspot cycles like the sun's, the result of steady thermonuclear furnaces at their centers. Still others may be so steady in their output that a spot only rarely mars the surface of a perfect disk. But we must consider a third possibility: stars whose sunspot phenomena become a dominant factor in the star's activity. Does the Milky Way contain stars that are scarred by a continual rash of giant starspots? The answer is yes.

## FLARES ON A GRAND SCALE

Compared with many other stars, the sun is a loner. About half of the stars in the Milky Way are found in double-, triple-, or higher-multiple-star systems in which two or more stars orbit their common center of mass. In many of these systems, interactions between stars produce unusual changes or conditions. One example of this is the RS CVn binary stars.

The abbreviation RS CVn stands for RS Canum Venaticorum, a type of variable star pair of which the first example was found in the constellation of Canes Venatici, the Hunting Dogs. The stars in this system vary their intensities, becoming periodically brighter, then dimmer over time. Although the RS designation simply refers to a listing of variable stars, for this system the RS might also stand for "really strange," because these are unusual even for variable stars.

What makes these stellar pairs so strange is the fact that each individual star in the pair strongly influences the behavior of its companion. The result is a star system with highly pockmarked stars that can eject flares of hot gas far beyond anything similar that our sun can do.

How do we recognize the RS CVn binary star systems? They are too distant, and their component stars too close together, to be seen as individual points of light. Instead, astronomers discover such stellar pairs through the study of their spectra—starlight spread into its component colors, or wavelengths.

When astronomers observe two stars close together in a binary system, the motion of each star around their common center of mass causes a continuously changing shift in the wavelengths of the spectral lines at which especially large amounts of light are absorbed or emitted. This shift arises from the Doppler effect, which changes all the wavelengths of the light emitted by a source moving toward or away from the observer. By observing these shifts, astronomers can easily determine which spectral lines come from which star in a binary system. By timing the cyclical changes in wavelength, they can discover how long it takes each star to orbit the center of mass.

The stranger aspects of these RS CVn stars are revealed, however, by their spectra. RS CVn binaries show unusual spectral lines produced by hydrogen and carbon. In our own star we see these lines only in the spectra of the light from the gas above sunspots.

This gas is hotter than other outer regions of our sun, and the spectral lines from hydrogen atoms there show large velocities that astronomers associate with the ejection of a flare. Similar spectral lines in RS CVn stars likewise reveal high-temperature gas in rapid motion.

But RS CVn stars also show features quite different from those seen in the sun. Astronomers record the behavior of the light from a star system on what they call a light curve—a graph of brightness measured over time. When they do this for RS CVn stars, they find a "dip"; that is, for a short period of time (days or hours), the light from the stars decreases. That's no surprise; it could easily be explained as the result of one star eclipsing its companion, blocking its light from our view. But if this were so, then the regularity of the stars' orbits would cause the dip to appear at regular intervals of time. RS CVn binary systems do show a regular dip, but in addition, unlike other binary star systems, they reveal a "traveling dip," a secondary dimming, whose appearance varies in time compared with the steadily periodic main dip (see Figure 11).

To an astronomer, the existence of a traveling dip suggests that one of the stars is rotating, and that the star has a brighter and a darker side. Rotation is no surprise—most stars do indeed rotate; our own sun does so in about thirty days. But the sun does not have a dark side. How then can stars in RS CVn systems achieve this effect? To produce traveling dips like those observed, one of the companion stars must have a surprisingly large fraction of its total surface area—perhaps 50 percent—far darker than the rest, and the dark area must change its location on the star as time passes.

Starspots—the stellar equivalents of sunspots—offer an answer, but one that raises more questions. The dark side could be a remarkable series of starspots, but the star would need perhaps a thousand times more spots than the sun has on a typical day. Could a star be poxed to this

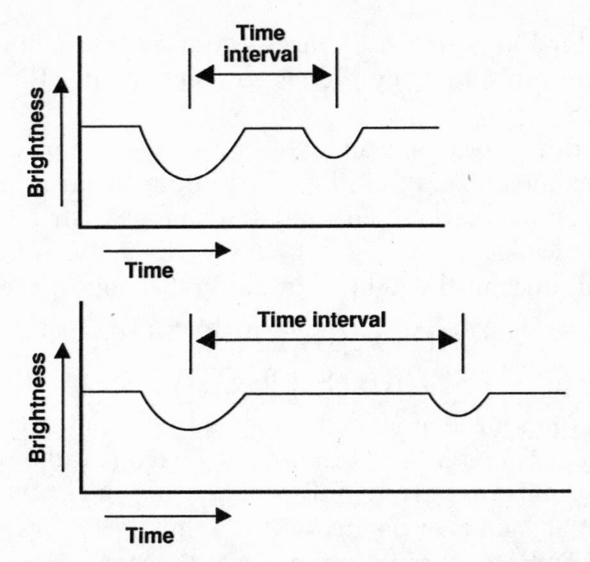


Figure 11: The varying brightness or light curve of an RS CVn star shows a "traveling dip"—that is, a decrease in brightness that changes its position along the light-curve cycle as time passes. *Drawing by Crystal Stevenson*

extent? And how can we check this hypothesis? Since astronomers can't actually obtain images of such troubled stars, can they use other tools of stellar forensics to clinch the case?

Surprisingly, they can. Stars like the sun emit not only visible light but also x-rays from the corona of million-degree gas that envelops the solar globe. The x-ray emission increases greatly whenever solar flares erupt. Hence an x-ray astronomer can tell when the sun has become more active: Solar x-rays arrive in abundance, and these x-rays are associated with the existence of sunspots.

RS CVn stars also produce x-rays. There is nothing odd about that: Many stars are found to have x-ray emission. But very few stars have a rapidly varying x-ray light curve and few change their x-ray intensity over a time scale of hours. Yet RS CVn stars do indeed change their x-ray intensity over a day or less, as if huge stellar flares intermittently enhanced the x-ray emission. The x-rays from RS CVn stars provide a second piece of critical evidence for flares in those stars—flares far more powerful than the sun's, for they produce a much greater flux of x-rays.

The third piece of circumstantial evidence for flares in RS CVn systems is found in the telltale spectral lines of hydrogen and carbon. For they too change their intensity in time scales of just a few hours or days, much like the spectral lines in the light from individual solar flares.

## RADIO REVEALS THE SECRETS OF STARS

One final piece of evidence seems to clinch the case for enormous flares in RS CVn binary systems: radio waves. These radio waves from a distant star are the latest chapter in the saga that begins with the man who first (accidentally) detected radio waves from the sun.

In 1898, in a desolate valley in Colorado shadowed by Pike's Peak, a Croatian-born American engineer named Nikola Tesla, a former technician for Thomas Edison, built the first wireless system to transmit and receive radio waves over any considerable distance. Tesla, a genius who pioneered the fluorescent light, alternating current, and dozens of other inventions, had a goal: the wireless transmission of power across the Earth using low-frequency radio waves. But Tesla faced a crucial problem: His crude receiving system was plagued by static. A student of lightning, Tesla knew that even distant strikes of natural electricity produced large amounts of radio static, static that he hoped to produce artificially and to modulate with his dumbbell-shaped transmitter/antenna.

Tesla could never pin down what produced the static detected in the absence of lightning, though he speculated that the sun might be partly responsible—an insight forty years ahead of astronomical confirmation. Tesla's plans for power transmission via radio waves remained a dream. Also he cannot be credited with the discovery of radio emission from the sun. That credit goes to the British radio engineers whose early-warning radar system detected radio emission from a solar flare in February 1942.

Today radio astronomers secure highly detailed images of the sun's radio emission with relatively little effort. These radio images reveal a complex picture of an immensely hot corona, within which solar flares rise from eruptions closer to the sun's surface. The intense heat produces x-rays (see Photo 11), but the radio waves arise from the streaming of electrons through the sun's magnetic field. It is exactly this process that provides the energy that powers our civilization. Electrical generators use huge rotating coils of wire surrounded by powerful magnets. The electrons in the coils, analogous to the electrons expelled from the solar surface, move at high velocities through the magnetic field. In nature, this causes a current to flow. Electrical generators confine most of the current they produce to wires for ease of use. This flow nonetheless amounts to a radio wave at a low (60 Hz) frequency. Thus moving electrons in a magnetic field produce radio power, both in the sun and in electrical generators.

What works for the sun likewise works in other stars. By now, almost a hundred stars have been discovered to produce some type of radio emission. But RS CVn systems are especially active emitters of radio waves. Most of them show evidence for a continuous "quiescent" radio emission—weak, but always there. Often, however, the radio emission from these stars will increase to a strength from ten to a thousand times greater than this quiescent strength in a time scale of a few minutes or hours.

This behavior is reminiscent of solar flares, which produce a temporary and fleeting spurt of radio static from the sun. But the sun produces nothing of this magnitude. To resemble an RS CVn star, the sun would need flares hundreds of times more powerful than the strongest ever recorded—flares we should hope we never experience.

The best studies of flares in RS CVn systems have been made with radio interferometers—groups of radio antennas linked together by computer to function as pieces of a single large dish (see Photo 12). Radio interferometers provide much more detailed maps than any single dish can. Even with radio interferometry, an RS CVn star in its quiescent phase looks like a pinpoint without detail. But a huge flare changes this: Astronomers can then see details, even though fuzzy or crude, that prove definitively that RS CVn stars are flarers *par excellence*.

Viewed through its radio emission, a flaring RS CVn star grows from a pinpoint to an expanding blob. Slowly, a fingerlike projection of radio emission protrudes, then fades over a few hours, as the blob dissipates like expanding fireworks. Hours later the phenomenon repeats. Days later things return to relative normality. The RS CVn star becomes a radio pinpoint again.

The details of the radio flares may not be clear, but the sizes are. Each one of these flares may be five thousand times the size of the Earth or more, a sudden protrusion of hot hydrogen gas boiling menacingly outward from a poxed star.

The cause of such powerful flares is the star's nearby companion, which disturbs the star into emitting a stream of hot gas by raising tides, unbalancing the star's outer layers and causing waves. The turbulence upsets the star's delicate balance of heat distribution and leads to the turmoiled spots.

Some of the matter ejected in a flare extends out to distances greater than the distance from the sun to Venus. This equals the distance between the star and its companion, which for its closeness gets a blast of the blowtorch it helped set off. Any planet around one of these stars would suffer the same fate as a neglected steak on a barbecue and be charred beyond easy recognition.

So honor the constant sun, with only an occasional sprinkling of sunspots, for the steady energy output that has allowed life to evolve and to flourish on Earth. And admire the RS CVn stars, whose rash of starspots can cover half the star, and whose flares, induced by a companion star, dwarf anything our sun can produce. Their erratic, awesome fires, mere pinpoints to us, help remind us of the vast distances that make for safety in the Milky Way.

Photo 1: The "milky way" consists of enormous numbers of stars, which we can see as individual points of light with the aid of a telescope. *Lick Observatory*

Photo 2: This spiral galaxy, seen nearly edge-on, resembles the Milky Way in its overall structure. If the sun belonged to this galaxy, its position would be most of the way out toward the edge of the visible galaxy. *Hale Observatories*

Photo 3: The Andromeda galaxy, two million light-years from our own, closely resembles the Milky Way and offers the best chance to examine another spiral galaxy. *Lick Observatory*

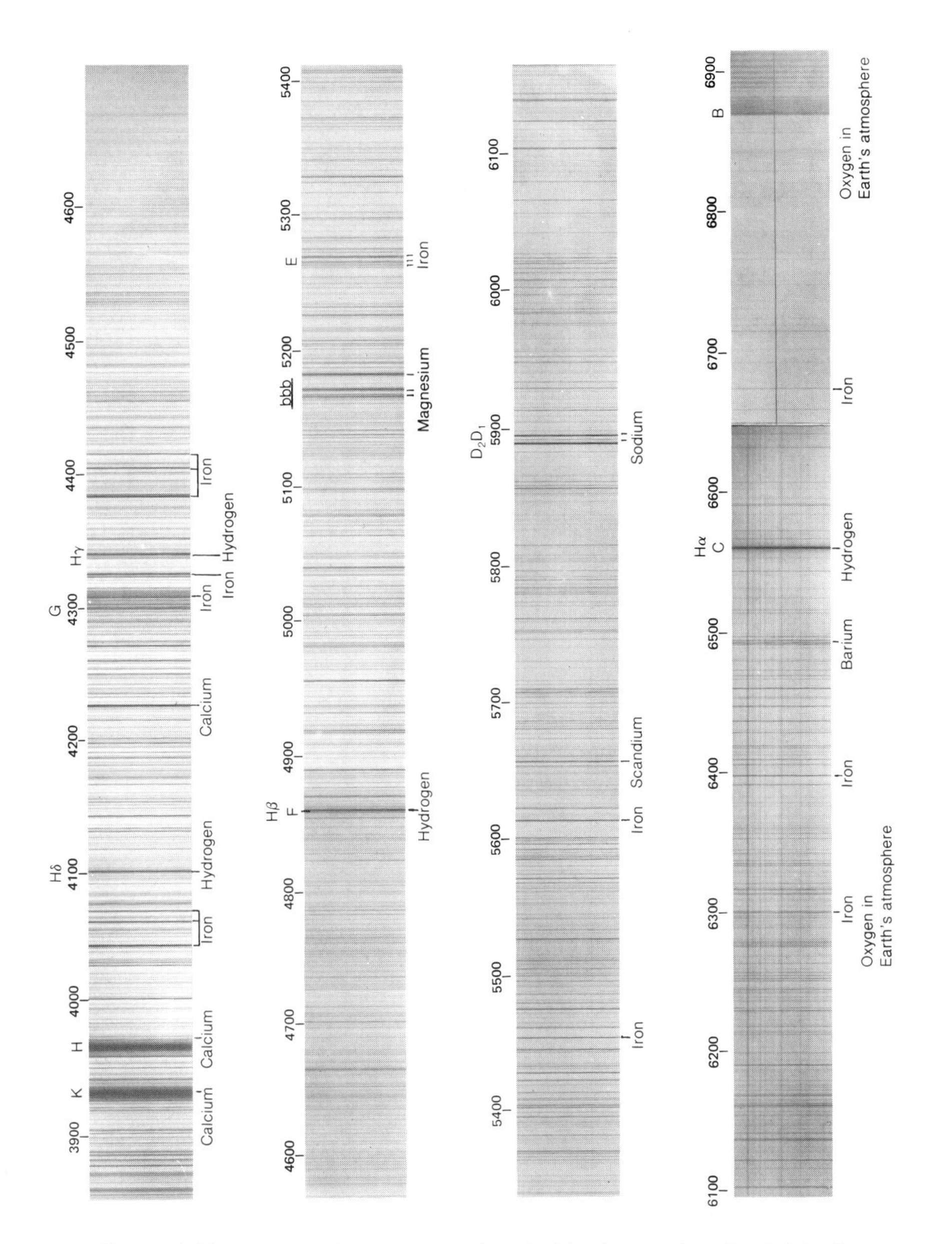


Photo 4: Each type of atom or molecule blocks or absorbs light of certain particular frequencies or wavelengths, producing a dark absorption line in the spectrum that shows all visible-light frequencies and wavelengths. This high-resolution photograph of the solar spectrum identifies some of the more prominent absorption lines by the type of element producing them; the numbers above the spectrum give the wavelength in angstroms (one angstrom equals $10^{-8}$ centimeter). By studying the "fingerprint" of these spectral absorption lines, as in the spectrum of sunlight shown here, astronomers can determine which types of atoms and molecules—and how many of each type—exist in a star's outer layers, where the blockage of light occurs. *Hale Observatories*

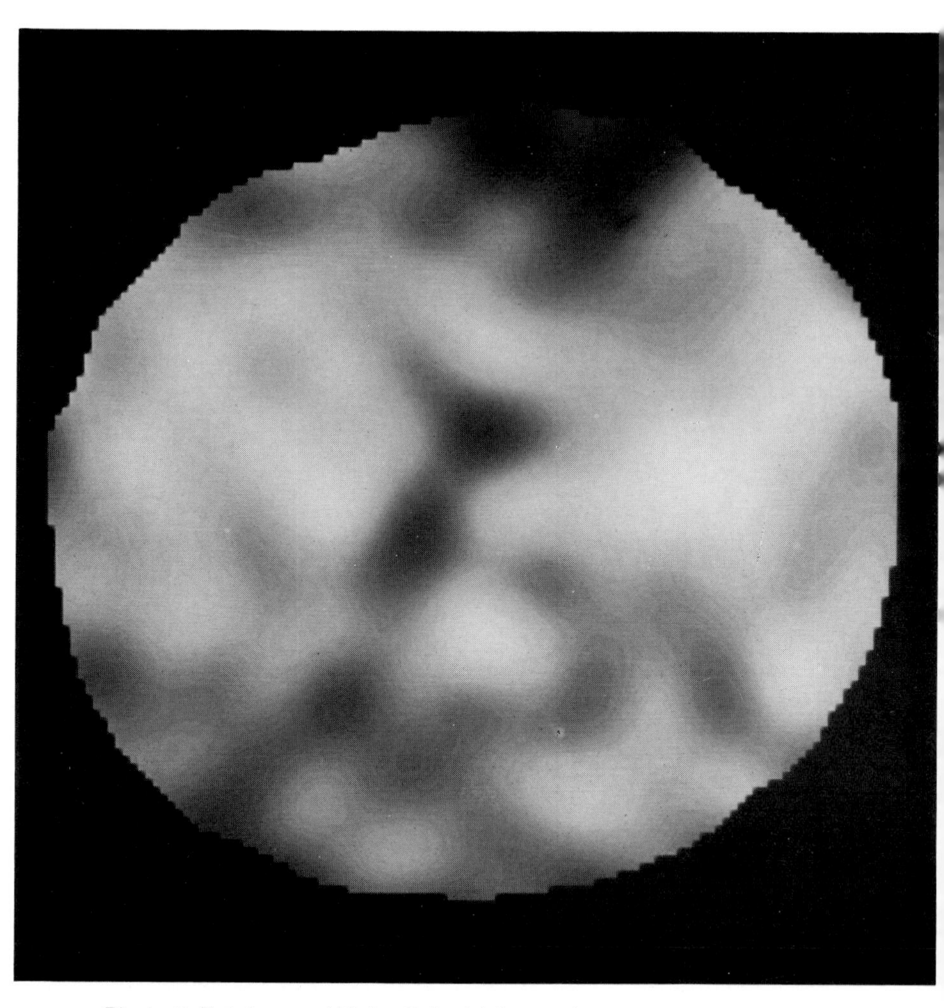

Photo 5: Betelgeuse (Alpha Orionis) is a red supergiant star, so large (and relatively close to the sun) that it is one of the few stars that can be seen as more than a single point of light. However, most of the apparent detail seen in this photograph is spurious. *Photograph courtesy of National Optical Astronomy Observatories*

Photo 6: A planetary nebula consists of the outer layers of an aging star, gently expelled during the star's red-giant phase. The star's core will (in most cases) eventually become a white dwarf. *Hale Observatories*

Photo 7: The visible Orion Nebula is a cloud of gas and dust some twenty light-years across, lit from within by young, hot stars that have recently formed within the cloud. *Lick Observatory*

Photo 8: The Orion Nebula forms part of Orion's "Sword," which appears between the three bright stars in Orion's "Belt" (top) and the even brighter star Rigel (lower right), which represents one of Orion's feet. *Yerkes Observatory*

Photo 9: In the center of the Orion Nebula, a new star cluster called the Trapezium lies half-hidden in its cocoon of dust and gas. *Lick Observatory*

# 6

# The Center of the Fire

Within each giant spiral galaxy lies a seething, violent core, a brilliantly lit maelstrom of closely packed stars and fast-moving gas revolving about a hidden, dominant center. Calculations allow astronomers to speculate about conditions there. They show that in the center of a spiral galaxy such as the Milky Way, stellar collisions, almost nonexistent elsewhere, occur frequently. Many stars explode as supernovae, producing interstellar shock waves that in turn affect nearby stars in the densely populated neighborhood. Still other stars merge to form seething, gigantic caldrons, each with the mass of hundreds of suns. Surrounding these immensely luminous giant stars, gas streamers as wide as a thousand solar systems pour into, or are shot out from—who knows what? A giant black hole?

With all the gas and dust about, the inner cores of galaxies are shrouded from our view. But these violent secrets are beginning to yield to astronomers' new techniques for observation. The closest of these centers of mystery lies some thirty thousand light-years away in the core of the Milky Way.

We might think that our position halfway toward the rim of the Milky Way offers a good view of our galaxy's core. But a clear night's view with the unaided eye proves that we cannot see anything of the center of our galaxy. Dust grains in the disk of the Milky Way absorb all the light from whatever lies at the galactic core (see Photo 13). To probe this region, we must abandon the visible in favor of other types of radiation.

## INFRARED AND RADIO OBSERVATIONS OF THE GALACTIC CORE

Even a good-sized optical telescope will reveal little information about the galactic center. The one thing that a telescope will show is the center's approximate location on the sky. Counting stars on long-exposure photographs reveals that the constellation Sagittarius houses the core of the Milky Way, for there the number of stars per square degree rises to a maximum. But these photographs show only stars and wisps of glowing gas, veiled in part by other bands of opaque gas. There is no clear, cosmic "x marks the spot." In visible light, the center remains hidden in a veil of absorbing material.

The first chance to pierce that veil arose with a new technique in astronomy: radio astronomy. In 1932 a radio engineer named Karl Jansky, an employee of the Bell Telephone Laboratories, founded radio astronomy by accident (see Photo 14). Using antennas he built from salvaged materials, and observing at wavelengths close to those of amateur radio shortwave bands, Jansky discovered that an annoying amount of the static on transcontinental radiotelephone conversations was generated not by cars or thunderstorms but by the universe itself. Of all sources of this "cosmic static," the galactic center was the strongest. But Jansky's antenna had so little direction-finding ability that he could not pinpoint the source of the radio emission precisely and thus could not verify that it indeed arose in the galactic center.

Astronomers were slow to exploit the new opportunities afforded by Jansky's discovery. Only a dozen or so years later, after World War II had ended, was Jansky's new method of observation pursued by them in earnest. At the much shorter radio wavelengths observed by the first professional radio astronomers, the galactic center indeed emerged as a radio source, one among many, named Sagittarius A (or Sag A) to indicate that this was the most intense radio source in the constellation Sagittarius. Only during the 1960s, more than thirty years after Jansky's discovery, did the advent of large dish antennas allow any reasonable detail to be observed in the radio emission from Sagittarius A.

The early radio images, maps of the radio waves from Sag A, revealed a complex, roughly egg-shaped region of radio-emitting gas. The radio waves arose not because the region was hot but rather because ionized hydrogen atoms had released their electrons, which were traveling through magnetic fields at nearly the speed of light. Astronomers found that radio waves produced in this way, called synchrotron radiation because it was first observed in particle accelerators called synchrotrons, also produce most of the radio emission from other powerful radio sources.

An unusual aspect of the galactic core was that different radio wavelengths probed the region to different depths. Some of the outer gas was opaque to some wavelengths of radio waves and transparent to others. Thus, by observing at different wavelengths, astronomers could record the center's details as if they were peeling an onion of layers differing in their radio opacity. Other radio sources tend to be too small for this technique to work. Even for the galactic center, this method of observing at different wavelengths in order to decipher the layers required special types of high-resolution interferometers that became available only during the 1970s. In the year 1970 Sag A was an intriguing but confusing object of study, still basically a mystery to astronomers.

While radio astronomy offered steady progress in the development of interferometers, another young field of astronomy, infrared astronomy, helped to answer some questions about the galactic center. Infrared astronomy emerged in the 1960s as an adjunct to optical astronomy, with infrared detectors mounted on existing visible-light telescopes. Infrared astronomy relies on the fact that the invisible wavelengths of radiation longer than those of visible light (but still short in comparison with radio waves) carry off most of the heat from cooler objects. Infrared observations are particularly good at revealing regions rich in molecules.

Most stars emit the bulk of their energy in radiation with the wavelengths of visible light and ultraviolet. But cooler stars (those with surface temperatures below three thousand degrees Celsius) should emit most of their energy as infrared radiation. Thus infrared astronomy offers an excellent means to find the cooler stars and gas clouds in the universe (see Photo A).

Infrared technology has proven a stubborn nut to crack. Only within the past decade has infrared astronomy had detectors able to secure detailed images of the sky. Even today, many infrared astronomers perform the simple but valuable task of measuring the amount of infrared radiation from various locations in the sky. This technique shows how much cooler material—stars or gas—exists at each location. Because infrared suffers far less absorption by interstellar dust grains than visible light, it offers a fine chance to probe the heart of our galaxy. For the region of the galactic center, these measurements immediately revealed that the galactic center, including the radio source Sag A, produces vast amounts of infrared radiation. The intensity of this infrared radiation increases markedly as astronomers look closer to the position of the galactic center. This strongly suggests that the density of stars—especially of cool stars—increases dramatically in regions closer to the center. Infrared observations first established that the Milky Way's core is a place where stars crowd together.

The next dramatic discovery about the center of the galaxy came from improved techniques in radio astronomy. During the late 1970s, radio astronomers trained their telescopes toward the galactic center, using radio spectrometers to look for "radio lines" (radio emission at specific wavelengths). These radio lines are caused by hydrogen atoms, by the "OH radical" (oxygen and hydrogen atoms joined in pairs), by carbon monoxide molecules, and by other molecules and atoms that produce radio emission at particular wavelengths. Molecules can form only in cooler, denser environments. They therefore typically appear in dense, dark clouds, such as those where young stars are forming (see Chapter 4). So radio emission from interstellar molecules may indicate sites of forming stars. But what radio astronomers really value about the radio lines is the opportunity they offer to discover the speeds and directions of motion of the regions from which they arise.

Radio astronomers have long known that radio lines allow the chance to infer the speeds of the gas producing these lines because the Doppler effect will shift the wavelengths of lines to shorter or longer wavelengths if the emitting gas moves toward or away from an observer. Radio spectroscopy that measured any shifts in wavelength thus provided an ideal method to measure the motions of the gas near the galactic center.

When radio astronomers first applied this method to the galactic center in the mid-1970s, they found that molecular clouds at temperatures a few hundred degrees above absolute zero form a ring around the periphery of the galactic center. These molecule-rich clouds take only a few million years to orbit the center. The discovery of this ring of molecular clouds in near-circular motion provided a crucial indication that our Milky Way's center much resembles the centers of other galaxies.

The development of infrared astronomy has kept pace with radio astronomy. Using state-of-the-art infrared detectors (similar to those that the military uses to detect the heat plumes of missiles and jet aircraft) infrared astronomy has made great progress during the past decade. One highly important observing tool has been NASA's *Kuiper Airborne Observatory*, a C-141 cargo jet specially modified to house a 36-inch infrared telescope and special instruments that compensate for the aircraft's motion. High above most of the atmosphere, the *Kuiper Airborne Observatory*, named in honor of an early pioneer of infrared astronomy, can avoid the greatest problem that confronts infrared astronomy—that, except for the shortest wavelengths, most of the infrared radiation from cosmic sources is absorbed by water-vapor molecules in the Earth's atmosphere and never reaches the surface. The *Kuiper Airborne Observatory* can therefore observe the galactic center in types of radiation invisible from ground-based observatories.

The *Kuiper Airborne Observatory* has allowed infrared astronomy finally to make detailed spectroscopic observations at key infrared wavelengths. These observations have revealed the hot gas at the galactic center, whose presence is revealed by an infrared spectral line of oxygen. Because this oxygen line arises only when gas containing the oxygen has been heated to many thousand degrees, infrared astronomers know that regions in which this line emission arises must have these high temperatures.

Most recently, radio astronomy has provided one additional improvement in our understanding of the Milky Way. During the 1980s, radio interferometers observed even finer detail near the galactic center. In the new observations, what previously looked like a smeared-out blob now showed fine detail: arcs and streamers, bands and rings, and nearly empty regions (see Photo 15). The technique of very long baseline interferometry (VLBI), which links radio telescopes around the world by computer to form a single giant interferometer, revealed a pointlike "star" at the center of Sag A. This pointlike radio source seemed (to astronomers) likely to arise from an "accretion disk" of matter spiraling into a black hole. To put this discovery in context, let us review what we now know about the regions close to the center of the Milky Way.

## OUTSIDE LOOKING IN

Close to the core of our galaxy, different phenomena occur at different distances from the center itself. We may therefore "peel the onion" and examine the galaxy's central regions, ordered by decreasing distance from the center.

*One Hundred Light-Years*: Here lies the galactic center's most remarkable structure: vast streamers of hot ionized gas that trail along the center's magnetic field, forming a lace of gas that is probably being drawn into the galactic center by gravity. The gas takes hairpin turns while it falls into the central heart, creating a dramatic pinwheel effect. Two distinct arcs of gas lie on opposite sides of the center.

*Twenty Light-Years*: The streaming gas clouds rich in carbon monoxide, hydrogen cyanide, and other molecules reveal the increasingly rapid motion and rising temperatures and densities as one approaches the core. A dumbbell-shaped double cloud of radio-emitting matter projects from the center, resembling the double radio sources seen in other galaxies.

*Five Light-Years*: The arcs form two parabolic structures of gas and stars. Even closer to the center, these structures join to form a ring of gas, which narrows to a bar of gas that crosses the ring.

*Two Light-Years*: Inside the ring, all the gas has fallen into the center. Something at the center exerts tidal forces that break stars apart and pull their matter inward.

*One Light-Year*: There is little structure visible here. But the region emits ultraviolet and infrared. Infrared spectroscopy shows that the gas is orbiting at enormous speeds. Miniblobs of a double radio source appear.

*The Center*: Here lies a single object, Sag A. Sag A's radio structure changes with time, sometimes presenting a pointlike structure, sometimes a more extended one. Sag A also varies over a few months' time in its radio intensity. It may be the site of gamma-ray emission, seen only once by gamma-ray-detecting satellites.

In summary, we can be sure that near the galactic center hot gas is being drawn inward, influenced as it moves by the strong magnetic field, until it falls into a starlike core. Bands of gas in rings orbit a region where gas is drawn in and disappears.

# THE HEART OF DARKNESS

Astronomers who speculate about the galactic core rely on what they know about the laws of nature, plus a few crucial observational facts. Among the latter is the measurement, using the Doppler effect, of the speeds within the ring of orbiting gas. From these speeds, astronomers can determine how much mass this object contains. The velocities at which the gas clouds move are high—several hundred kilometers per second. From this speed, and the distance of the orbiting gas from the center, one can calculate the amount of mass needed to hold the ring of gas in orbit. This mass turns out to be several million times the mass of the sun.

In other words, more than a million times the mass of the sun must exist within a few light-years of the galactic center. Compare that with the mass within a few light-years of our sun: Instead of one or two stars with the sun's mass, pack in several million. Even if each of those stars has ten times the mass of the sun, there is still a packing problem: How do all those stars fit into such a small space?

In fact, once a million stars occupy only a few cubic light-years of space, their mutual gravitational attraction and frequent collisions will soon make them coalesce. The theoretical impossibility of maintaining stars as individuals corresponds to the lack of evidence that any large groups of stars exist near the galactic center. Instead, observations reveal only gas—and evidence suggestive of a single pointlike object drawing gas inward.

These facts lead astronomers toward an exciting possibility—the galactic center contains a massive black hole. Such a black hole could pack several million suns' worth of mass into a region much smaller than the solar system. The black hole would consume any gas that came near it, forming a rotating accretion disk within which gas heats to enormous temperatures before it falls all the way in. From the outside, all that would be visible is the accretion disk; the black hole would remain hidden.

Astronomers now speculate that every giant spiral galaxy has a supermassive black hole at its center. These black holes might have been the first part of the galaxy to form, and would have served as gravitational seeds that attracted much more matter into the giant gas clouds that became protogalaxies. This scenario has yet to be verified by direct observation, but it offers a tempting way to help resolve one of the great mysteries of astronomy: how galaxies ever managed to form within the initially smooth and homogeneous early universe.

If galaxies do contain supermassive black holes with accretion disks of hot matter orbiting around them, we may expect that, as sufficient time passes, the regions around these black holes will cease to emit much radiation. The accretion disks require "food" in the form of matter falling inward to join the disk of matter spiraling inward. A galaxy must then eventually exhaust its food supply, leaving the supermassive black hole with nothing around it. Billions of years from now, astronomers on planets orbiting the slowly dying stars that circle the galactic center may find it a far less intriguing region for their investigations. For the present, we seek to continue our attempts to understand what goes on in the heart of darkness at the center of the Milky Way.

# 7

# JETSTARS

STARS MAY BE spheres, but not every celestial object is spherical. Objects in the universe show a variety of shapes: round planets (some with rings), tailed comets, wispy cosmic gas and dust clouds, ringed nebulae, pinwheel-shaped spiral galaxies, and so on. But none of the shapes on this list describes the largest single entities in the universe. These are the double radio sources, galaxies with huge clouds of radio emission that dwarf the visible galaxies, sometimes by a factor of a hundred or more. Stretching over distances greater than a million light-years, these radio-emitting regions resemble twin turbulent gas clouds, typically forming dumbbell-like shapes with the visible galaxy (when it *is* visible) in the center (see Photo 16).

These double radio sources present astronomers with a puzzle. Their radio emission arises from the synchrotron process, when electrons accelerated to nearly the speed of light move through magnetic fields. However, in view of the rate at which the radio sources emit energy, they should disappear in a few million years as their electrons slow down and cease to produce radiation. Somehow new electrons are continually being accelerated to nearly the speed of light, or by now we should observe almost none of the double radio sources.

With the advent of high-resolution radio interferometers during the late 1970s, part of the answer became clear: The electrons are produced in jets that are shot out in opposite directions from the center of a galaxy. Remarkably narrow and highly directional, the jets move outward at speeds close to the speed of light. When the jets strike the highly rarefied gas that permeates intergalactic space, the fast-moving electrons lose their highly directional motion and form vast clouds of radio-emitting gas.

Cosmic jets have ranked among the hottest topics of astronomical research in recent years because they are powerful and energetic—and because astronomers strive to understand where they come from. Why should a galaxy eject matter at such tremendous speeds in two narrow jets? And why don't we see such jets in our Milky Way?

## SPINARS

Even before the jets were revealed by radio interferometers, their existence was predicted by astronomers. The theory was that large spinning stars (or perhaps black holes) called spinars attract gas but shed some of it as the gas falls into them. The ejection of gas occurs primarily because the infalling gas is strongly heated and also starts to rotate along with the spinning object. The pressure of the radiation on the hot gas can combine with the centrifugal force on the rotating gas to fling some of it outward. Thus a spinar resembles an enormous rotating gyroscope from which hot gas is ejected. The bulk of the infalling gas creates an accretion disk of matter orbiting the center. From the spinar, gas can escape only through small holes where the spinning disk lets some matter get through. The spinar's accretion disk forms a tube (perhaps reinforced by the guiding effect of the star's magnetic field) that keeps the escaping gas directed, like a stream of water traveling in a hose (see Figure 12). By the time the gas emerges it has formed two thin streams traveling at nearly the speed of light in two opposing directions.

Even the best gyroscopes are unstable. Precession, or wobble, will make the pole of rotation of a gyroscope point

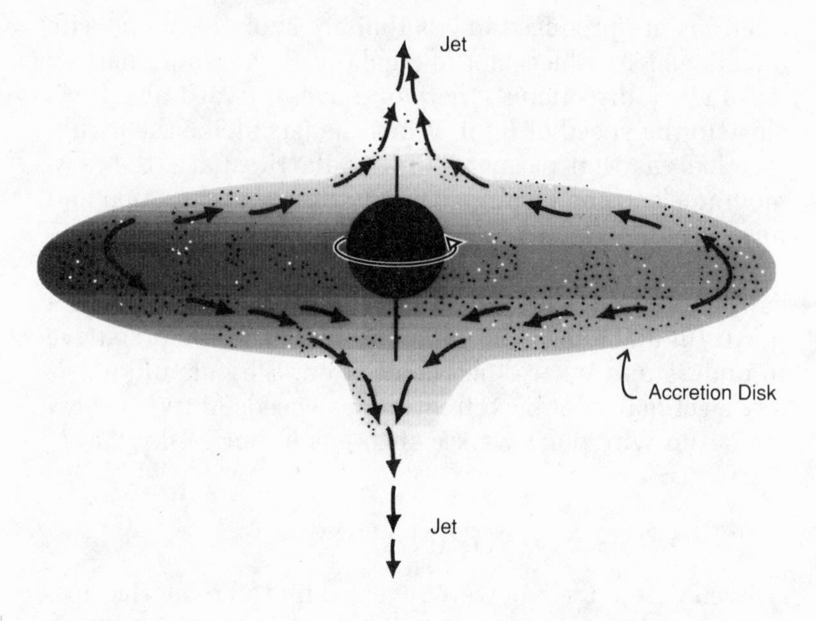


Figure 12: The spinar model for the production of jets in radio galaxies includes a rotating accretion disk of matter from which jets emerge in opposite directions. *Drawing by Crystal Stevenson*

in different directions. Similarly, if a spinar's axis of rotation changes its direction, the spinar's jets will wobble in space, projecting over a fanlike pattern as time passes. Some double radio sources certainly look as though their jets of ejected matter have been spread into more unusual shapes by slow changes in the direction of the jets. But most double radio sources look as if they arise from spinars that have been stabilized against such wobbling, and have been ejecting these jets in steady streams for millions of years.

Whatever causes the spinar phenomenon, it is not rare; astronomers have found thousands of galaxies in which double radio sources prevail. A few galaxies even exhibit jets that may be observed in visible light, spikes of emission in their galactic pinwheels. The Milky Way's center (see Chapter 6) seems to contain a small-scale double radio source, although jets have not yet been observed.

Our galaxy may not itself be a giant double radio source, but since the mid-1970s astronomers have found jets in our own Milky Way on a smaller scale. Radio jets in the Milky Way appear in two different forms, arising from two different types of stars. Here we shall explore the basic types of jets in our galaxy: those in very young stars and those in parasitic stellar pairs.

## THE YOUNG STARS

Young stars with jets, discovered only during the past few years, are related to an important phenomenon called water masers, often found in the molecular clouds that give birth to young stars (see Chapter 4). At radio frequencies of about 22 and 183 gigahertz—well into the microwave portion of the radio spectrum—some of these molecular clouds emit intense radio lines from molecules of water. These lines are highly confined in wavelength and exhibit other odd characteristics: Their emission is polarized (like the light passed by some sunglasses); this emission arises from astronomically tiny regions, varies by factors of ten over a time span of weeks, and reveals velocities (measured by the Doppler effect) quite different from the velocities of the molecular clouds within which the emission arises. To astronomers, these characteristics imply that the water molecules in these clouds are subject to processes that make them emit intense streams of radio radiation in certain directions. These processes are similar to those that produce laser beams that shine with incredible brightness in preferential directions.

Some of these water masers in molecular clouds are the strongest radio sources on the sky at their frequencies of emission. In fact, with a modest microwave radio receiver and a small dish antenna (even smaller than those used to detect television signals from satellites) the narrow-channel hiss of a few of the most intense water masers would become audible. Cosmic water masers provide excellent natural navigational aids because they are so strong and their positions are so precisely known.

To the radio astronomer, water masers posed a problem: Why aren't they moving at the same velocities as their parent clouds? The problem is that in some clouds many separate water masers are seen to move at different velocities, and the range of velocities can be several hundred kilometers per second, as is the case in the giant molecular cloud called W49. If that range of velocities really indicates motion, then parts of these molecular clouds appear to be exploding.

In 1980, radio interferometry showed that water masers have another kind of velocity in addition to the motions revealed by the Doppler effect (toward or away from the observer). Precise but minute changes in the sky positions of the water masers revealed that they also have enormous velocities *across* our line of sight. These water masers are some of the fastest moving objects in the Milky Way!

For the most part, water masers in a molecular cloud have no preferential direction of motion; they seem to be moving randomly outward from an extrapolated starting point in the molecular cloud. And there at the starting point lay collections of young, massive stars. A simple conclusion follows: The young stars emit such intense radiation that it pushes away the cloudlets that contain the water masers. Over time, the radiation pressure accelerates the cloudlets to several hundred kilometers per second.

In some of the molecular clouds, the water masers turned out to have locations grouped by velocity. The water masers with motions toward us lay mostly on one side, while those receding from us lay mostly on the other. This correlation of location with velocity was difficult to explain as the result of an explosion, or from being pushed away by radiation pressure; in these cases cloudlets should be moving away in all directions. The resolution of this puzzle helped to point the way to locating newly formed stars.

During the early 1980s, astronomers observed some of these smaller molecular clouds with radio spectrometers, searching for spectral lines produced by carbon monoxide, hydrogen cyanide, and other types of molecules. Some of these molecular clouds contained water masers, but others didn't. What the radio astronomers found was that observations of water masers provided only part of the picture. They found that the other spectral lines traced out the motions of the entire molecular cloud, not just the bits and pieces—the cloudlets that contained water masers. From the new studies, a picture developed of huge flows of gas emerging in two opposing directions from a young star or stars. For the most part the motions were not so extreme as those of the water masers, but the correlation of speed and position first seen for the water masers was confirmed and clarified.

In molecular clouds young stars appear to be spewing gas from their vicinity mainly in two directions. The radio maps of these regions resemble their double radio source counterparts—except for the jets. Instead of a strikingly narrow shape, the jets are more like wide funnels pushing gas in two directions. Because these are "jets" only in the broadest sense, these molecular cloud bubbles were named bipolar outflows. Bipolar implies both the ejection of material in opposite directions and the specific method that astronomers believe to have created these bubbles: The young stars apparently have their own accretion disks. Two methods are possible to eject matter from these disks. On the one hand, gas falling toward the star can be shot outward by the force of radiation pressure. In this case, we should see gas shooting out in two funnel-shaped regions above and below the accretion disk. On the other hand, the accretion disk could break into bits, which might later form planets.

These two methods could produce broad funnels of ejected matter. But just as a thunderstorm isn't a tornado, a funnel is not a jet. To see true jets emerging from young stars we must examine a class of objects called T Tauri stars. These T Tauri stars, named after the first member of the class to be studied in detail, vary their visible-light intensities—a characteristic of either quite young or aging stars—and show broad spectral lines. We now know why those lines are broad: They arise in the accretion disk surrounding a T Tauri star. The velocities observed in these lines indicate that the star is in the process of shedding some of the matter in its accretion disk. The shedding leads to bipolar flows—and to jets. Some photographs of T Tauri stars show the material emerging from them (see Photo 17). At the end of the line of ejected material, astronomers find . . . water masers! These particular water masers are a special subclass of the water-maser phenomenon, called Herbig-Haro objects after the two astronomers who discovered them, George Herbig and Guillermo Haro. Zipping away from the T Tauri stars at speeds of several hundred kilometers per second, the Herbig-Haro objects are cosmic Roman candles of gas shot out by jets from the accretion disk around a young star. A Roman candle—a type of fireworks that ejects balls of fire—provides a good model for Herbig-Haro objects since the jets don't necessarily produce a continuous flow. Just why T Tauri stars have such well-directed jets remains a mystery. But they do provide us with cosmic jets of a juvenile kind (see Photo B).

## PARASITIC STAR PAIRS

T Tauri stars have the same general properties as the spinars that power double radio sources in other galaxies. But the spinars in double radio sources produce jets that move outward at speeds ten to a hundred times greater than the mere few hundred kilometers per second for the jets found near T Tauri stars. The Milky Way contains at least one amazing object whose properties may be more representative of the centers of radio-emitting galaxies. That object has the unassuming name SS433.

SS433 is number 433 in a list of blue starlike objects compiled by Bruce Stephenson and Nicholas Sanduleak. In 1979, examination of the object's spectrum presented astronomers with a startling discovery. Like other stars, SS433 exhibits many strong spectral lines produced by hydrogen atoms. But SS433 has *two sets* of hydrogen spectral lines. Furthermore, the wavelengths of both sets of lines changed markedly with time, and the changes repeated in a cyclical manner every five and a half months (see Figure 13). No astronomer had ever seen anything like this before!

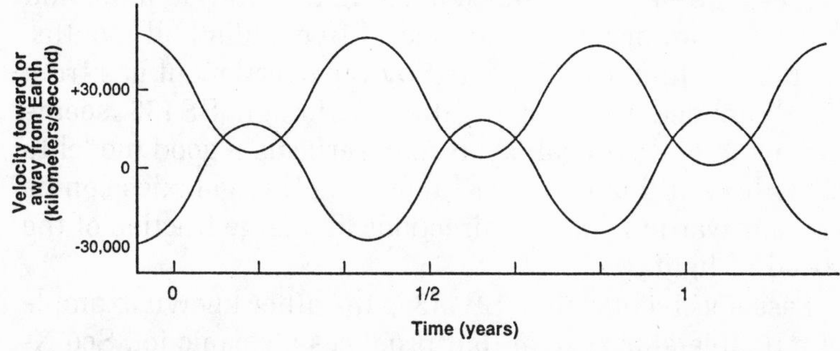


Figure 13: The strange object SS433 shows tremendous changes in the frequencies and wavelengths at which it emits much of its radiation. The graph shows these changes, which probably arise from the Doppler effect and suggest the existence of beams of material shot out at a large fraction of the speed of light. *Drawing by Crystal Stevenson*

After months of testing and discarding other possibilities, only one explanation seemed reasonable. The two sets of spectral lines must arise in two jets, each moving at more than one-quarter the speed of light. In addition, the axis along which the twin jets emerge in opposite directions must be wobbling in space, with a five-and-a-half-month period. As this wobble proceeds, the two sets of lines will reverse their deviations in wavelength from the average value, as first one jet and then the other points more toward us than away from us.

SS433 also proved to be a radio source, and a spectacular one at that. Observed with very long baseline interferometry, SS433 not only showed evident jets, but even their wobble was visible. The observations revealed a corkscrew of material pushed outward by the jets, a corkscrew that blended into a huge duck-shaped mass of gas called W50 that contains SS433 within it.

Today, the pattern for SS433's behavior can best be explained with a model in which an accretion disk orbits a collapsed object called a neutron star (see Chapter 8). Of crucial importance in the model, this accretion disk is continually replenished by matter from a bloated companion star, so the accretion disk never runs out of fuel. The accretion disk shoots out jets along its rotation axis, and this axis wobbles with a period of five and a half months. SS433 is a kind of spinar, fed by the overflow of gas from its distended companion star. Although SS433 seems unique in our own galaxy, it may serve as a good model to show how still more powerful jets in other galaxies can be shot outward in specific directions at a large fraction of the speed of light.

Less spectacular than SS433 is the other known example of a double-star system that produces a cosmic jet, Sco X-1. The most intense x-ray source in the constellation Scorpius, Sco X-1 is likewise believed to be a neutron star with a stellar companion. But that companion furnishes little gas, at least at the present time. Sco X-1's accretion disk is probably sparser, and its ejected gas radiates less intensely than SS433's. Its jets are not visible.

But radio observations of Sco X-1 reveal something akin to T Tauri stars and Herbig-Haro objects. Blobs of radio-emitting gas shot from Sco X-1 move outward in opposing directions. We can't measure the velocities of this gas by using the Doppler effect because we can't see spectral lines as we do in SS433. But with radio interferometry, we can detect the motion of the radio-emitting regions across our line of sight. Perhaps with more sensitive observations we will be able to see the jets directly. But since many double radio sources have no jets detected, we must assume that some jets are invisible and can be seen only when they slow down and fragment into bubbles of hot gas.

Stars with jets thus arise at two key stages in a star's evolution: as the star forms, as is the case with T Tauri stars and Herbig-Haro objects, and during the late stages of a double-star system, as in Sco X-1, where a swollen, aging star spills its outer layers onto a collapsed companion. In order to understand this latter phenomenon, we must turn to one of the key moments in this history of stars, and ask what happens to a star when it runs out of fuel with which to shine.

# 8

# NEUTRON STARS AND PULSARS: BEWARE THE BLACK WIDOW

THE COLLAPSE OF a star's center that begins a supernova explosion can leave behind a highly compressed object, one of the strangest beasts of the cosmos—a neutron star. A neutron star is a stellar core that consists of neutrons packed to an incredibly high density within a small volume. The concept of a neutron star originated with two European-born astronomers in Pasadena, California, Walter Baade of the Mount Wilson Observatory and Fritz Zwicky of the California Institute of Technology (Caltech). In 1934 these two scientists suggested that within an exploding star, the stellar core would collapse and turn into neutrons.

Baade's and Zwicky's idea lay unexplored for a few years. Then in 1938 the brilliant physicist J. Robert Oppenheimer, working with a graduate student and another physicist, asked the question: Does nuclear physics say what would happen if you compressed a starlike mass to a smaller and smaller size? The answer, Oppenheimer saw, was that you might get what we now call a black hole. But if you pressed not quite so hard that you produced a black hole—the ultimate gravitational sink of matter—you might make the star turn all its heavy particles into neutrons, the elementary particles that along with protons form the nuclei of all atoms. If the star turned all its matter into neutrons, it would truly deserve the name neutron star. Five decades after Oppenheimer's work these neutron stars, which were once merely the creations of theorists speculating with equations, have come to be recognized as an important class of objects in the universe.

A neutron star seems complex at first glance, but as astronomical oddities go, the neutron-star concept is more easily understood than that of a white dwarf, despite the fact that astronomers had found dozens of white dwarfs long before they suspected that neutron stars might be abundant. A white dwarf is the dying core of a sunlike star, containing both electrons and atomic nuclei (chiefly carbon nuclei). White dwarfs' relatively complex composition makes their internal structure more difficult to comprehend than that of neutron stars. A neutron star consists of neutrons all the way down.

## MAKING A NEUTRON STAR

Consider a star whose center collapses because it has no way to support itself against its own gravitation. Deprived of any chance to liberate energy through nuclear fusion, the individual nuclei in the star (mostly iron nuclei at the time that the collapse occurs) fall toward the center, propelled by their mutual attraction. As described in Chapter 12, this tremendous inrush requires only about a second to occur—one second in which the star's core shrinks from a size larger than the Earth to one comparable to the island of Manhattan. The violence of this collapse hurls the nuclei against one another at enormous speeds, and their collisions break the nuclei apart into protons and neutrons. But this is not all: We must not forget the electrons in the star.

Electrons play no role in conventional nuclear fusion. This is because nuclear fusion proceeds through strong forces, the attractive forces that hold protons and neutrons together in atomic nuclei. But electrons are entirely immune to strong forces. To attempt to make electrons fuse with other particles by using strong forces is therefore like attempting to glue water to a board; it simply won't happen. The fact that electrons feel no effect from strong forces means that they play little role in most of a star's basic operation, which consists of fusing smaller nuclei into larger ones. But electrons do feel weak forces, which cause reactions to occur among elementary particles that are more subtle, and typically much slower, than those that arise from strong forces. The fact that electrons feel weak forces changes the entire star—once it becomes sufficiently small and dense.

Weak forces are *much* weaker than strong forces. It is hard to provide sufficient vividness to this "*much*" because the factor involved is about ten trillion ($10^{13}$). This means that when nuclei (which feel both strong and weak forces) fuse together, the strong forces do most of the work, but one part in ten trillion of whatever occurs must be attributed to weak forces. We can easily conclude that if both strong and weak forces operate in a particular situation, we are unlikely to see the effect of weak forces easily. To do so, we need a situation where strong forces do not operate at all and we can "look on beauty bare," as Keats put it.

Electrons offer such a chance. Because electrons do not respond at all to strong forces, when we see an electron fuse with another particle we are seeing only weak forces at work. But since weak forces *are* so weak, we will never see this occur in most astronomical situations—not on Earth, nor between the stars, nor even in the centers of ordinary stars. To make electrons fuse with other particles solely under the influence of weak forces requires a highly unusual astronomical situation. We need matter so dense that the electrons' ordinary reluctance to fuse—the result of weak forces' basic weakness—becomes of no account in view of the tremendous pressure put upon the electrons that makes them fuse.

The question of how much density will make this fusion occur has a definite answer: about ten trillion times the density of water. With such an enormous density, even though the electrons are still nowhere near "shoulder to shoulder" with the nuclei, they will nevertheless fuse with them, rammed together by repeated violent collisions between electrons and nuclei. The key reaction is the fusion of an electron with a proton, producing a neutron and a neutrino. In a collapsing stellar core, this reaction occurs about $10^{57}$—one billion trillion trillion trillion trillion—times. The collapse breaks the nuclei into neutrons and protons, and then nearly *every* electron in the collapsing core finds a proton and fuses with it.

The result is a large number of neutrons and a nearly equal number of neutrinos. The neutrons stay around but the neutrinos do not: They leave the scene at nearly the speed of light, relatively unimpeded by the matter at enormous density, and depart in all directions, carrying with them the news that another stellar core has bitten the gravitational dust and collapsed. As described in Chapter 12, Supernova 1987A announced its appearance on the scene with just such a burst of neutrinos—a burst that unfortunately went unnoticed until after the later pulse of light had attracted astronomers' attention.

So the neutrinos rush out into space, but the neutrons stay behind and make a neutron star—an object with the mass of a star but only a dozen or so miles across, made almost entirely of neutrons. All of these neutrons combine to form what amounts to a single giant nucleus: Just as a helium nucleus consists of four particles—two protons and two neutrons—held together by strong forces, a neutron star consists of $10^{57}$ or so neutrons, bound by gravity and held in a single nucleus by strong forces. Note that the neutron star in one important respect is a *simpler* object than the helium nucleus: The latter consists of two different types of particles, whereas a neutron star has only one, neutrons.

And so, as Oppenheimer perceived, a stellar collapse can produce a single giant nucleus, as large as a New York borough but far more massive. The neutron star has a mass comparable to the *sun's* mass, 330,000 times the mass of the Earth. This means that matter in the neutron star must be packed to fantastic densities, ten to a hundred trillion times the density of water.

This density is *enormous*. Yet it is just the density of "nuclear matter"—the matter in atomic nuclei—throughout the universe. Every molecule in our bodies contains nuclei whose density equals that within a neutron star. To be sure, these nuclei have diameters less than a trillionth of a centimeter, but they fundamentally resemble the matter in a neutron star (though our nuclei are a bit more complex). This helps to remind us that we consist mainly of empty space, the vast and lonely reaches (on an atomic scale of sizes) in which electrons orbit the tiny, amazingly dense nuclei of their atoms. If our bodies were packed as solidly as the atomic nuclei they contain, our masses would increase by ten trillion times, and each of us would have more mass than Mount Everest.

## SUPPORTING NEUTRON STARS AGAINST GRAVITY

Each neutron star packs a star's mass into a small volume. It also performs the amazing feat that Oppenheimer predicted: It keeps from collapsing despite its tremendous density, which produces enormous self-gravitational forces—forces seemingly capable of making anything collapse.

What holds up neutrons stars against the incredible crushing fist of their own gravity is called the "exclusion principle." The exclusion principle describes the fact that particles of certain types—neutrons, for instance—cannot be packed more tightly than a certain critical density. Though the principle is not a type of force, it can act in ways reminiscent of forces. In particular, the exclusion principle can prevent an object's further contraction under its own gravity by the simple refusal to let the object's constituent particles pack together any more closely. Physicists can pack neutrons as closely as they like (provided they have sufficient force to do so) until the density reaches a certain critical value; then they can never pack them any more tightly, no matter how much force they apply.

That critical density equals about a hundred trillion times the density of water—not surprisingly, the density of nuclear matter and the matter that comprises a neutron star. Nature has so arranged itself that an object made from neutrons will be bound together by its own gravitation and will assume a density equal to that set by the exclusion principle. At this density a single teaspoonful of matter, if brought to the Earth's surface, would weigh as much as a battleship! Luckily, we seem in no danger of meeting this dreadnought teaspoonful—but nature has its places where this type of matter exists by the Manhattanful.

Those places are the neutron stars: made of neutrons, denser than one can easily claim to imagine, held together by gravity, and held up by the exclusion principle. If this were all, then neutron stars would be just another oddity of nature, which we could pass by with a tip of the hat to Oppenheimer and a sigh of amazement at yet another beast in the cosmic zoo. But there is far more to neutron stars, thanks to two of their less fundamental properties that we have yet to consider. Neutron stars rotate at high speeds, and they have enormously strong magnetic fields. Taken together, these properties are responsible for pulsars. And pulsars are responsible for many astronomical careers.

## PULSARS FROM NEUTRON STARS

Astronomers say that "a rapidly rotating, highly magnetized neutron star produces a pulsar." What they mean is that if you take a tremendously strong magnet a few miles across, spin it many times per second, and surround it with a cloud of electrically charged particles, the spinning magnet will accelerate some of the charged particles and make them emit streams of electromagnetic radiation—radio, infrared, visible light, and even x-rays. The rest is detail; significant detail, to be sure, but detail nonetheless.

That is the good side: We think we understand the basic mechanism that makes a pulsar work, and we think we know the broad outlines of the process that produces such a pulsar. The dark side is that we lack a real, in-depth understanding of why a pulsar *pulses*—that is, emits its radiation not as a continuous stream, the way stars do, but in a punctuation of greater and lesser amounts of radiation, repeating with incredible accuracy.

## THE DISCOVERY OF PULSARS

More than two decades ago, news from England stood the world of astronomy on its ear: Radio astronomers at Cambridge University had found a source of radio waves that produced pulses of emission spaced at precisely repeating intervals of 1.337 seconds. This first pulsar—to use the name soon coined by Thomas Gold of Cornell University—had been found by Jocelyn Bell, a graduate student at Cambridge, as she monitored a new sort of radio antenna, which consisted simply of wires strung at regular intervals on wooden posts. The antenna, coupled to a radio detector, was capable of detecting radio waves of many centimeters' wavelength, and it detected best those objects that passed directly overhead. Since the Earth's rotation carried many objects across the sky, Cambridge University thus had a low-cost radio antenna capable of detecting longer wavelengths than were usually studied, and of observing those portions of the sky that passed near the zenith (the overhead point) as seen from central England.

The detector needed no human operator: A chart recorder passed a spool of paper at a constant rate beneath a pen, which was connected to the antenna and detector. The pen drew a short line for modest power reaching the antenna and a longer line for greater power. When Jocelyn

Bell examined the chart recorder's record one autumn morning in 1967, she noticed a "bit of scruff," a place where the recorder pen had moved significantly back and forth repeatedly for several minutes. She eventually convinced herself that this was not a glitch in the machine and reported her observations to her professor, Anthony Hewish. Two years later, Hewish shared in the Nobel Prize for the discovery of pulsars. Jocelyn Bell has stated that she feels no regret that she was not included, for her role was to note what the pen had recorded, whereas Hewish was the first to understand the importance of the regular, repeated pulses of radio waves. This may or may not be an appropriate distribution of rewards in this world; at any rate, it is a familiar one.

The first pulsar was named CP 1919+21 (CP for Cambridge pulsar; 1919 for the astronomical coordinate called right ascension (19 hours 19 minutes); 21 for the coordinate called declination (+21 degrees, that is, 21 degrees north of the celestial equator on the sky). Within a year several more pulsars had been found; within another year, dozens more. During the late 1960s an astronomer had to contend with pulsars prefixed not only with CP but also with MP (for the Molonglo Observatory in Australia), AP (for the Arecibo Observatory in Puerto Rico), and so on. Finally astronomers saw the uselessness of recording the place of the pulsar's discovery in the astronomical name of the pulsar. They dropped the CPs and MPs and APs and now call pulsars simply by their astronomical coordinates.

## THE CRAB NEBULA PULSAR

Within a year after the first pulsar discovery, the most famous pulsar of all was found: the Crab Nebula pulsar. By this time speculation was already strong (though hardly universal among astronomers) that pulsars were fast-spinning neutron stars. Pulsars should therefore have arisen when neutron stars formed in the collapse of a stellar core to begin a supernova explosion. If this was so, a likely site to find a pulsar would be a place where a supernova has recently exploded. Astronomers searched the stars within the Crab Nebula, which was known to be the remnant of an exploding star, and soon found (after a few false starts) an object whose radio emission pulsed far more rapidly than any pulsar then known: thirty-three times per second. The rapid pulsation seemed reasonable because the Crab Nebula is the remnant of a star seen to explode less than a thousand years ago, in A.D. 1054. The other pulsars did not lie at the location of a known supernova remnant, though one more such pulsar was soon found: the Vela pulsar. If the other pulsars had no associated remnant to be seen, it was reasonable to conclude that these pulsars arose from older neutron stars. In such cases, the explosion that produced these neutron stars has long since dissipated the ejecta, the parts of the exploded star that remain visible, as is true for the Crab Nebula and the larger and older Vela remnant.

So everything seemed to fit together, and this became even more true when the Crab Nebula pulsar was found to be emitting pulses of radiation in visible light as well as in radio. In fact, the Crab Nebula pulsar, flashing on and off thirty-three times per second, had been photographed for years: It is one of the two rather faint objects visible in high-quality photographs of the nebula (see Photo 18). But no one had thought to photograph these objects with a high-speed camera capable of seeing whether or not they were flashing on and off thirty-three times per second. (To be honest, such cameras did not exist until the late 1950s, and until the pulsar period was known—in this case established from radio observations—there would have been little chance even with such a camera of finding the visible-light pulses from the Crab Nebula.) The visible-light pulses coincided exactly in time with the radio pulses. Clearly something about the neutron star was making the pulsar wink on and off at a steady rate.

## WHAT MAKES PULSARS PULSE?

The soon-to-emerge consensus was that the something could only be the neutron star's rotation. Although one can imagine other possible ways in which a neutron star (or any other astronomical object) could change its situation regularly—for example, by oscillating its shape between a more spherical and a less spherical configuration—only rotation offered the chance of maintaining such exact regularity as was soon measured for the Crab Nebula and other pulsars. A neutron star that was alternately more squashed and less squashed might go back and forth between the two modes fairly regularly, but within a year, calculations showed, its period of oscillation would change noticeably. The period of the Crab Nebula pulsar's emission changed by less than one part in a thousand over that time—and even that change, as we shall see, has a natural explanation if we assume that it arises from the neutron star's rotation.

So rotation seemed the logical answer to the question of why the pulsar should pulse so regularly. In addition, the neutron star's rotation offered a chance to understand the basic mechanism by which a pulsar *produces* its radiation. This chance has taken more than two decades to reach partial fruition, but today most astronomers will agree on the way that a rotating neutron star produces streams of radiation at many different wavelengths.

## HOW A PULSAR WORKS: SYNCHROTRON RADIATION

Consider a rotating magnet ten miles long (see Figure 14). The magnet exerts electromagnetic forces on any electrically charged particles in its vicinity, just as a laboratory magnet will line up iron filings on a piece of paper placed over the magnet. As the magnet spins, it tends to pull any charged particles along with it so that the parti-

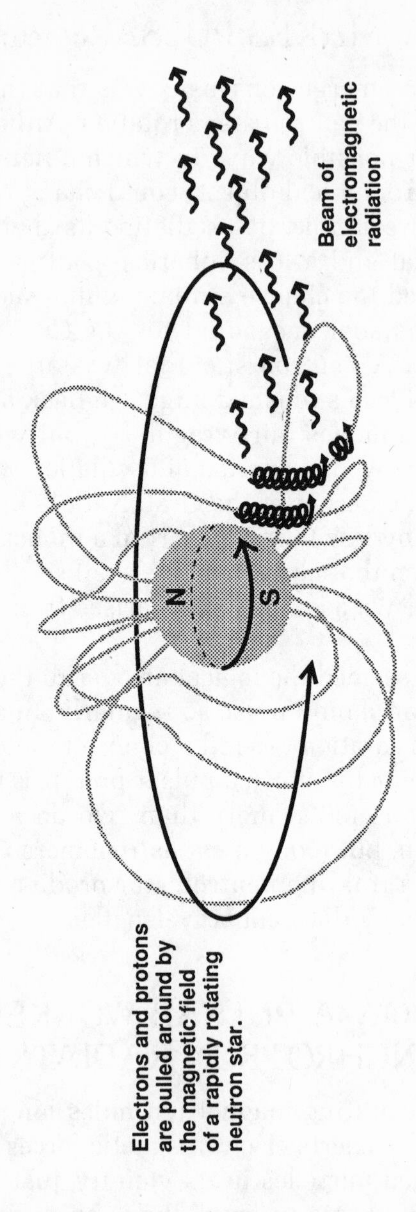


Figure 14: A pulsar resembles a giant rotating bar magnet, which sweeps up charged particles close to its surface and accelerates them to enormous velocities. "Hot spots" of especially intense radiation create a "lighthouse-beam" effect. *Drawing by Crystal Stevenson*

cles tend to move in circles along with the magnet. But the particles do not follow the magnet's motion perfectly because their inertia—their unwillingness to respond immediately to a force, which arises simply because the particles have mass—makes them lag behind the spinning magnet's changing forces upon them.

As a result, when the magnet spins, the nearby charged particles do move, but they do not simply circle around and around forever. Instead, as the particles orbit they tend to slip outward from the magnet. The net result is that many of the charged particles eventually escape from the environment in which they were moving. This escape has important consequences, but the immediate effect of the particles' rapid motion is this: They emit radiation because of the synchrotron process.

The synchrotron process draws its name from the class of particle accelerators called synchrotrons. Physicists who used these machines to create enormous magnetic fields, and then used these fields to accelerate charged particles to nearly the speed of light, noticed that an eerie glow emanated from the machines once in operation. Theorists explained this emission of visible light as follows: When charged particles move at nearly the speed of light in the presence of a magnetic field and change either their speed or their direction of motion, electromagnetic radiation will inevitably be produced. More radiation will be produced as the particles' speed increases, as the number of particles grows larger, as the magnetic field grows stronger, and as the particles undergo a greater acceleration (that is, change either their speed or their direction of motion by greater amounts). The act of emitting radiation takes energy from the particles, which must therefore slow down (and soon cease to produce radiation) unless some process resupplies energy to keep them moving at speeds of at least 99 percent of the speed of light.

The radiation produced by the synchrotron process was named synchrotron radiation. Synchrotron radiation can be recognized by its spectrum—the distribution of the numbers of photons in the radiation with different wavelengths or frequencies. Stars and other objects that emit radiation simply because they are hot show a characteristic spectrum in the photons they emit called the blackbody or ideal radiator spectrum. Such a spectrum, plotted as the number of photons with various frequencies against the photons' frequencies, rises slowly to a peak as the frequency increases and then declines sharply (see Figure 15). In contrast, the spectrum of synchrotron radiation, which does not arise from a region because it is hot, simply declines steadily: The number of photons at lower frequencies always exceeds the number at higher frequencies.

Thus once astronomers learned about synchrotron radiation they had a valuable tool to recognize explosive events across the universe. Since synchrotron radiation is typically strongest at the lowest frequencies—that is, in the radio domain of the electromagnetic spectrum—radio astronomers became those most familiar with the synchrotron process. During the past three decades they have identified region after region in the Milky Way, as well as in other galaxies, from which synchrotron radiation emerges in large amounts. *None of these regions is a star; rather, they are each the scene of an explosion.* Only an explosive process can accelerate particles to nearly the speed of light, one of the four basic requirements for synchrotron radiation. (The other three are magnetic fields, charged particles, and acceleration—requirements easier to satisfy than reaching speeds close to the speed of light.) Synchrotron radiation thus identifies explosive locales, and by studying the details of the synchrotron radiation, astronomers can learn a good deal about the conditions that exist in the region that produces it.

In a pulsar, the explosion has already occurred, and the synchrotron radiation persists because of what the explosion left behind: a rapidly rotating neutron star. The neutron star resembles a giant bar magnet whose magnetic field exerts forces on any charged particles left over from the explosion (typically protons and electrons) that sweep

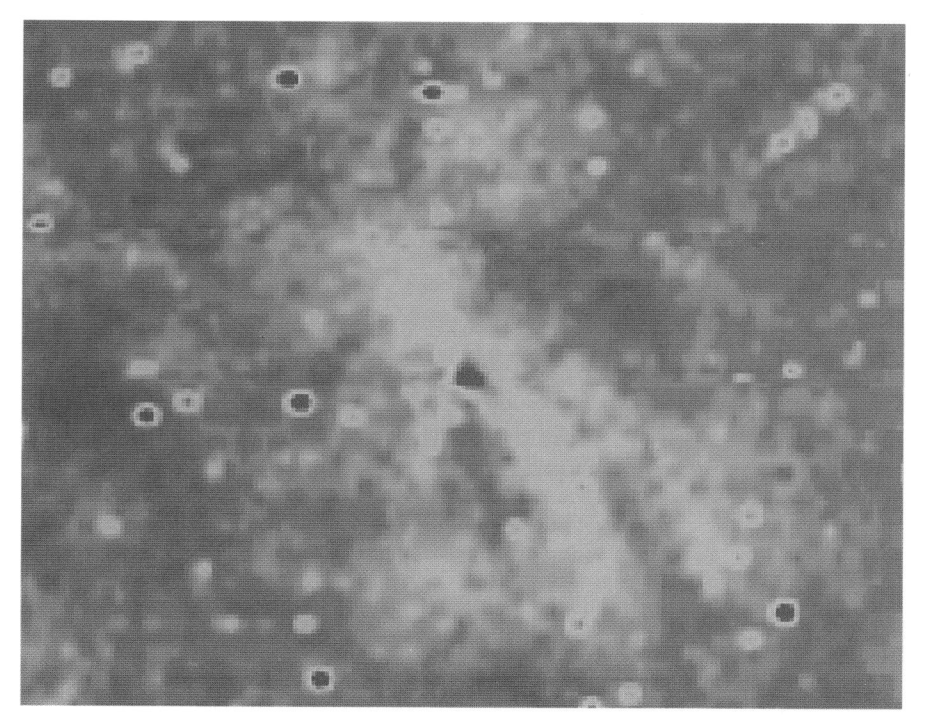

Photo A: An infrared map of the galactic center shows a host of objects emitting infrared radiation, which can penetrate the dust that blocks visible light. *Photograph courtesy of Prof. Gerry Neugebauer, copyright by the California Institute of Technology*

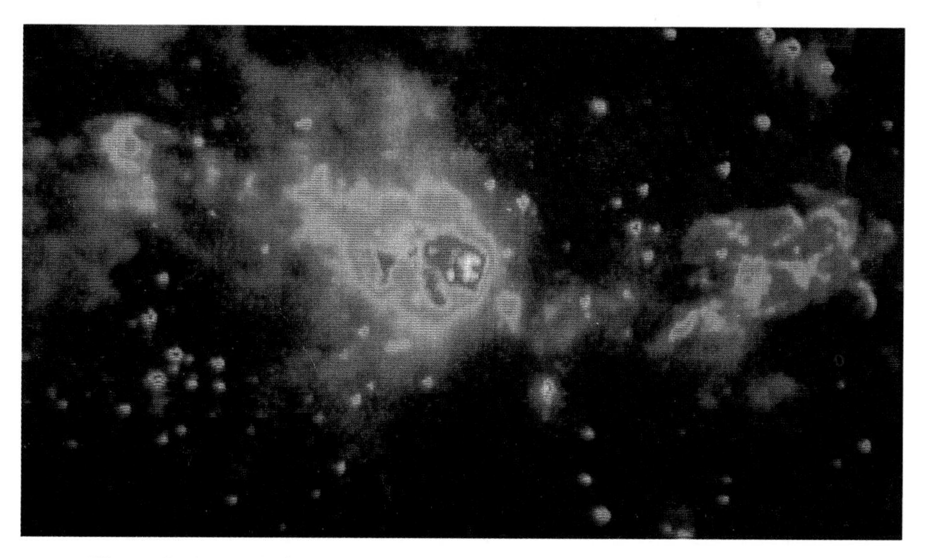

Photo B: A newly formed star has a bipolar outflow of gas along two opposite directions. This radio map shows the oppositely directed streams of gas. *Courtesy of Prof. Adair Lane, Boston University, and Dr. John Bally, AT&T Bell Laboratories*

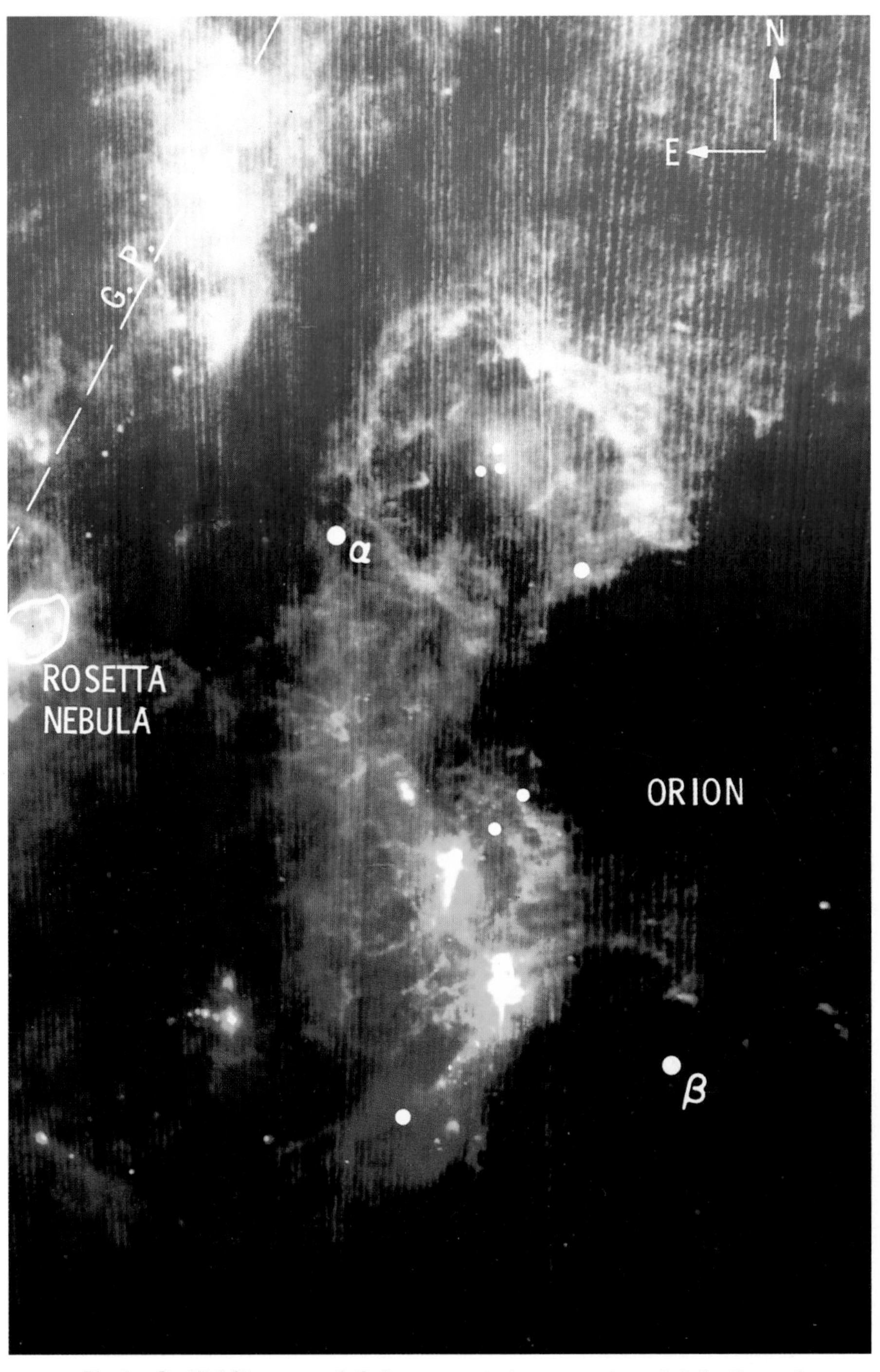


Photo C: *IRAS*'s map of Orion revealed a complex distribution of matter at what astronomers call "cool" temperatures, between "room temperature" and absolute zero. *NASA*

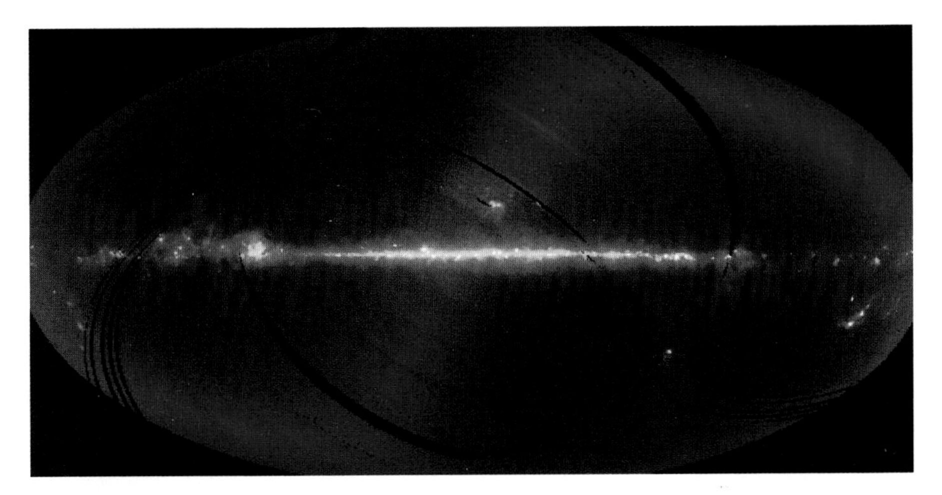

Photo D: The *Infrared Astronomy Satellite* (*IRAS*) surveyed the heavens in infrared wavelengths in 1983, producing this map of the sky, which lacks only a few swaths that the survey did not cover. *NASA*

Photo E: The prototype interferometer for the LIGO system consists of two perpendicular arms, each forty meters long, down which laser light shines back and forth many times. *Photograph by Alexander Abramovici, courtesy of the California Institute of Technology*

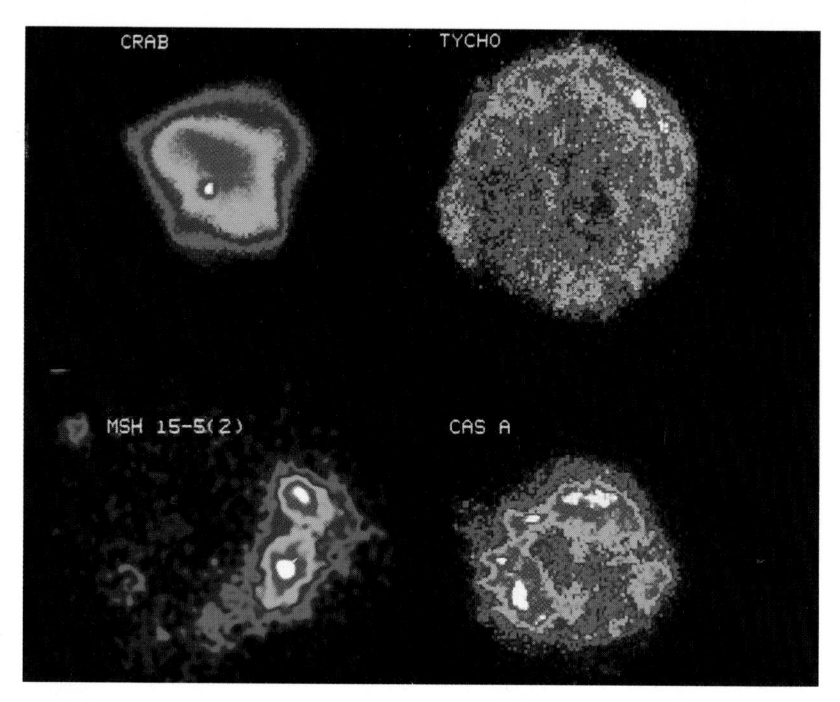


Photo F: Maps of the x-ray emission from four supernova remnants reveals roughly spherical shells of extremely hot gas ejected by the supernova explosions. *Courtesy of F. Seward, Harvard-Smithsonian Center for Astrophysics*

Photo G: This detector, containing some of the world's clearest water and a battery of complex equipment to record light flashes with precision, detected half a dozen of the neutrinos that Supernova 1987A sent through the Earth. *Photograph courtesy of Prof. Lawrence Sulak, Boston University*

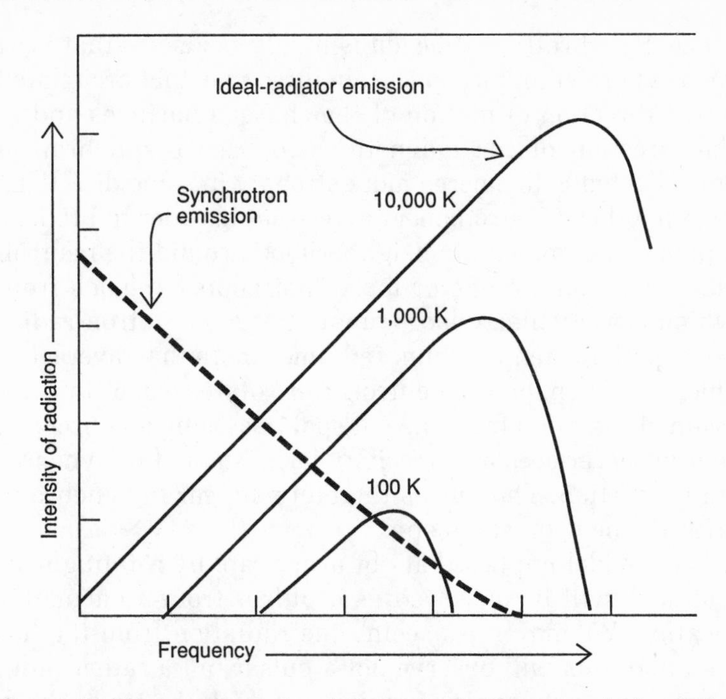


Figure 15: The spectrum of radiation—that is, the amount of radiation at different frequencies—for synchrotron radiation and the radiation from an ideal radiator differ markedly. Astronomers can recognize the process producing the radiation from the shape of the spectrum that is observed. *Drawing by Crystal Stevenson*

them into rapid motion (see Figure 14). Some of these charged particles reach nearly the speed of light, and since they are moving in the presence of a strong magnetic field and changing their direction of motion rapidly, the stage has been set for copious streams of synchrotron radiation to emerge from the vicinity of the neutron star. Furthermore, as the particles lose some energy through the very act of emitting synchrotron radiation, the neutron star's rotation, which drives the rotation of the magnetic field surrounding the neutron star, provides the ongoing energy supply needed to keep the synchrotron radiation flowing.

As yet, though, we have not described a pulsar but rather a source of steady synchrotron radiation. What makes a pulsar *pulse* remains a mystery in its details but is generally agreed upon (by pulsar theorists) in its essence. Synchrotron emission is highly beamed—that is, it emerges preferentially in certain directions that are related to the direction of motion of the charged particles and to the direction of the magnetic field. Hence synchrotron emission tends to emerge more strongly in some directions than in others. Astronomers agree that in order to produce a pulsar the immediate neighborhood around the neutron star must contain one or more "hot spots," regions from which a particularly large amount of synchrotron radiation—perhaps ten to a hundred times more than average—emerges. Then, as the neutron star rotates, each time the beam of radiation from the hot spot passes by an observer, that observer sees an especially large amount of synchrotron radiation. The pulses therefore appear in synchrony with the neutron star's spin.

This model implies that not every rapidly rotating neutron star will be observed *as a pulsar* from a particular location. We may miss seeing the radiation from the hot spot and thus will observe not a pulsar but a rather mild, nearly constant source of radiation, which we are likely to miss because a repetitive variation will be noticed far more readily than a rather steady, lower-level output.

The several thousand pulsars found so far in the Milky Way are all believed to follow the basic patterns outlined above. Most of them have pulse periods ranging from a modest fraction of a second up to a few seconds. Pulsars must eventually slow down and radiate less strongly because the energy lost in synchrotron radiation must be replaced by energy from the rotating neutron star. This robs the neutron star of rotational energy, and astronomers have verified that most pulsars very gradually lessen their rotation rates, with occasional glitches caused by star quakes—modest readjustments of the neutron star's surface as its rotation rate decreases. (The gravity on a neutron star is so strong that the surface is smoother than a billiard ball; any "mountain" as high as a millimeter would immediately be pulled flat by gravity.) A few million years after the supernova explosion that formed the neutron star and the pulsar, the pulsar should become so faint and so slow in pulsation that our present search techniques would probably not detect it.

## A NEW WRINKLE: THE MILLISECOND PULSARS

Until the 1980s all seemed well with pulsars: Theory matched observation, and the newest pulsars should be, collectively speaking, the fastest, as the Crab Nebula pulsar's thirty-three beats per second implied. Then a modest reassessment of theory became necessary when Don Backer and his collaborators at the University of California, Berkeley, discovered a new class of pulsars, the millisecond pulsars.

Millisecond pulsars pulse with periods of approximately one-thousandth of a second—that is, they pulse about one thousand times per second. One might think that such pulsars must be extremely young, and youth certainly represents one way to explain a rapidly rotating neutron star. If astronomers ever find a pulsar in Supernova 1987A, they expect it will be a millisecond pulsar. But newborn, near-millisecond pulsars must be extremely rare; even the Crab Nebula pulsar, only a thousand years old, has already slowed down to only thirty-three pulses per second.

Instead, the millisecond pulsars are believed to be *old* pulsars that belong to double-star systems. Material falling from the companion star onto the neutron star can have the effect of "spinning up" the pulsar—but only after some million years have passed. After all, it takes time to make an object with the sun's mass rotate a few thousand times more rapidly than before. The first detected millisecond pulsar pulses 642 times per second, and astronomers believe that it has gone through a phase in which it slowed its rotation and pulsed at a rate slower than one pulse per second. Only later did sufficient matter flow onto it from its (hypothesized) companion to make the neutron star rotate first at tens of times per second, then at hundreds of times.

More than half of the nine millisecond pulsars now known appear in globular star clusters. Globular clusters rank among the oldest objects in the Milky Way (see Photo 19), so it is not surprising that they contain many of the most highly evolved pulsars.

A thousand pulses per second or one, hot spots or none, pulsars represent a triumph of astronomical theory and observation—a new class of objects that astronomers have fitted into their general scheme of stellar evolution. The millisecond pulsars represent a subclass of pulsars in which one star affects a close companion. Because billions of double- and multiple-star systems exist in the Milky Way, we should not be surprised to find that nature has devised a few special wrinkles on the general theme of one star changing another close by.

## BLACK WIDOW PULSARS

Among the most exotic (and exotically named) astronomical objects discovered in recent years are the black widow pulsars, of which only one representative has been found to date but which may exist in relatively large numbers in certain parts of the Milky Way.

The class of black widow pulsars owes its name to the fact that astronomers can deduce that each such pulsar must be steadily consuming a star in orbit around it. In 1988 the first (and so far only) black widow was found in the direction of the constellation Sagitta by a team of astronomers from Princeton University. The team was led by Andrew Fruchter and included Joseph Taylor, who has come to be recognized as the king of pulsars by most of his colleagues. Fruchter, Taylor, and their coworkers found a peculiar pulsar. This pulsar belongs to the class of millisecond pulsars, since it rotates 622 times per second. But unlike every other known pulsar, millisecond or not, this object's pulses are *eclipsed* for about fifty minutes at a time. These eclipses recur on a regular cycle, implying that the pulsar moves in orbit with a companion star that repeatedly eclipses it.

From these eclipses Fruchter, Taylor, and their colleagues deduced that the pulsar and a giant bloated object are moving in orbit around their common center of mass. This orbit has a nearly circular shape and a diameter of about four million miles—less than five percent of the distance from the Earth to the sun. Further calculations show that because the bloated object has a low mass (only a few percent of the mass of the sun) and has grown so distended, most of its volume lies within the pulsar's gravitational sphere of influence; that is, the pulsar has the greater claim, by gravity, to determining the direction in which the matter moves. The astronomers speculate that we may be observing the final stages of a pulsar "swallowing" its low-mass companion star. As we have seen, the fast-spinning millisecond pulsars owe their rapid rotation to the infall of matter from such companion stars. Reaching for biological analogies, the astronomers found the name black widow pulsar because a female black widow spider consumes the male with which she has just mated.

Soon after the radio observations had revealed the black widow pulsar, the same team of astronomers used the two-hundred-inch telescope on Palomar Mountain to obtain faint images of both the pulsar and its companion. They found that the pulsar's visible-light emission also undergoes eclipses, which correspond in time with the eclipses observed in the radio emission. The black widow pulsar therefore seems definitely established as a member of the galactic zoo. If one such object exists, with a pulsar swallowing a companion star, we may conclude that many others likewise exist but do not happen to have the proper orbital alignment for us to observe eclipses. Also, black widowhood clearly represents a temporary phenomenon: Once it has consumed its companion, the black widow pulsar runs out of fuel and can no longer call attention to itself through the eclipse phenomenon.

# 9
# The Search for Other Planetary Systems

Our sun orbits through the Milky Way with a swarm of smaller objects in gravitational tow: Nine planets, some sixty satellites of those planets, several thousand known asteroids, and an enormous number of meteoroids and comets all orbit our star, some of them along nearly circular paths, others in highly elliptical orbits. The legacy of Copernicus—the reluctance to assume that our position in the cosmos has a particular claim to fame—implies that many, perhaps most, of the other three hundred billion stars in the Milky Way should have similar attendant swarms of smaller objects. But can we verify this directly? Can we find the planets that we assume are in orbit around the sun's neighboring stars?

## HOW TO FIND PLANETS AROUND OTHER STARS

Suppose that you are convinced—or just plain interested—that planets may exist in orbit around other stars. How would you propose to find out for sure?

One possible method—excellent if it works—is the following: Assume that if enough planets exist around the three hundred billion stars in our galaxy, on some of them intelligent life will have developed. On some of those planets the intelligent life-forms will decide to explore the Milky Way for purposes of investigation, colonization, colonialism, or just plain good eating. Some of these exploring civilizations will eventually come our way. Thus the best way to discover planets around other stars is to wait for their inhabitants to arrive.

This method is beloved by those who sell hard-hitting magazines, those who read the magazines, and by UFO fans across the globe. It cannot be faulted as one way to proceed (unless you believe that it is chosen by every form of intelligent life in the galaxy, in which case everyone is waiting and no one is exploring). But this method leaves much to be desired by those who desire action; and it also has the drawback of relying on a long chain of deductions about what will happen on the (hypothetical) planets in our galaxy. A broader approach concentrates on the planets themselves rather than upon the life-forms that may or may not have evolved on the planets. An astronomer would be quite excited to discover extrasolar planets even before speculation began on the evolutionary history and present state of the planets around other stars.

Suppose then that you not only want to learn about extrasolar planets but are also willing to work at it. What sort of work should you do?

Well, of course you should *look* for those planets. You have fine telescopes at your disposal, but your search for the planets by direct means suffers from a tremendous handicap, one intimately related to the objects of your search. In astronomical terms planets nestle quite close to their parent stars. Combine this fact with an additional obvious point: Planets shine only weakly because they produce no visible light of their own and merely reflect starlight reaching their surfaces. The combination of the planet's proximity to its star and its low brightness rules out direct observation of extrasolar planets. Not even the best telescope operating at its theoretical best can reveal a planet the size of Jupiter orbiting one of the closest stars at a distance equal to Jupiter's distance from the sun (see Photo 20).

Since Jupiter is the largest of the sun's nine planets and

since the Jupiter-sun distance may well be typical of planets' distances from their stars, direct observation of extrasolar planets would require finding a much larger planet than Jupiter in orbit around a nearby star. Astronomers share this conclusion, at least so far as observations from the Earth's surface are concerned. Sealed within our bowl of protective but interfering atmosphere, even the best Earth-based telescopes can never hope to spot any planet comparable to Jupiter around any star beyond the sun. What we need is a new angle. We must either send a telescope into space (not a bad idea, discussed later in this chapter) or—a better bet for the short-term future—devise other, better ways to spot extrasolar planets than to see them by observation of the visible light that they reflect from their stars.

## INFRARED DETECTION OF PLANETS

One promising method of extrasolar planet detection lies in using the infrared region of the spectrum. Infrared radiation has longer wavelengths and smaller frequencies than those of visible light, and infrared is typically produced by objects cooler than stars. All of the sun's planets emit copious amounts of infrared radiation simply because they have temperatures of several hundred to a thousand degrees above absolute zero, even though they produce almost no visible light at all for the same reason—their temperatures. It takes surface temperatures of several thousand degrees Celsius (for red stars) up to tens of thousands of degrees (for blue stars) to make large amounts of visible light emerge from a hot object. So if planets produce infrared, why not look in infrared to spot them around faraway stars?

The problem is that stars produce plenty of infrared radiation too—far more than any planets around them. To be sure, a star like the sun emits far more energy each second in visible light than in infrared—but it emits huge amounts of infrared as well, thousands of times more than

Jupiter, the most infrared-luminous of its planets. Jupiter's temperature of about 120 degrees above absolute zero leaves it at a temperature disadvantage in comparison with the Earth (300 degrees above absolute zero) or Venus (1,000 degrees above absolute zero), but because Jupiter is so much larger than the Earth or Venus it nevertheless outshines the smaller planets in infrared emission. The trouble is that the sun outshines Jupiter by about ten thousand times.

Let's look on the bright side: In infrared radiation the sun outshines Jupiter by *only* a factor of ten thousand. In visible light the sun's brightness exceeds Jupiter's by a factor of one hundred million! In other words, to find a Jupiter around a nearby star is ten thousand times *less difficult* if you use infrared rather than visible-light radiation for your search.

Good news indeed. The difficulty then becomes that infrared-detecting technology is far less advanced than visible-light detection. The former approach commands less respect than the latter. As a result—and partly because the technology of infrared detection seems inherently more difficult—even a helpful factor of ten thousand won't solve the problem. With present-day technology we can't hope to find planets around other stars by their infrared emission.

But the future looks better. If we develop somewhat better infrared detectors—an improvement by a mere factor of ten to a hundred should do—we could hope to find a Jupiter-like planet in orbit around Alpha Centauri at a Jupiter-like distance (that is, at a distance from Alpha Centauri comparable to Jupiter's distance from the sun). So we must simply proceed to instruct our infrared-detector technologists to move forward with their plans in order that within a few years we may have a system, to be mounted on a fine telescope such as the new Keck Telescope at the Mauna Kea Observatory in Hawaii, that may reveal the first known extrasolar planets.

# PLANET DETECTION FROM THE SPACE TELESCOPE

On April 24, 1990, carried aloft by the Space Shuttle *Discovery* and carefully set adrift at an altitude of 375 miles above the Earth's surface, the mightiest, most expensive instrument ever built for astronomy finally began to observe the heavens. The Hubble Space Telescope, named in honor of the man who found that the universe is expanding, provides us with the first true observatory to orbit above the atmosphere, free from the atmospheric blurring and absorption that rob all Earthbound telescopes of complete clarity of view (see Photo 21).

One of the most longed-for advances to be furnished to astronomers by the Space Telescope was the potential to *see planets*. The orbiting observatory has a coronagraphic finger—a narrow rod that projects into the field of view, covering up the star itself but leaving the regions around the star accessible to close scrutiny. On Earth this coronagraphic technique can be used—with great care and a clear sky—to observe the inner regions of the sun's corona of hot gas, which surrounds the sphere of the sun but which usually disappears from view because the light from the solar surface far outshines the corona. The shimmering, refracting atmosphere of Earth makes coronagraphic observations difficult for the sun and impossible in the search for planets around stars, since the weak reflected light from any such planets will surely be washed out in the light that is bent—thanks to our atmosphere—around the sides of the coronagraphic finger.

The Space Telescope was designed to profit from the fact that no atmosphere exists in space to bend light around the coronagraphic finger. This fact gave the Space Telescope a good chance to find any Jupiter-like planets that might orbit around the few dozen closest stars. But the news from space in the summer of 1990 changed all this: The Space Telescope's mirror turned out to have been imperfectly made, robbing the telescope of the chance to see the cosmos as clearly as it was meant to. Only if and when a correcting optical system is installed can the Space Telescope begin its efforts to discover extrasolar planets. Even when this occurs—now thought possible sometime in 1993—the detection of extrasolar planets around the closest stars will still lie just at the edge of what the Space Telescope should be able to do. Hence it may turn out, with regrets all around, that even the Space Telescope will be unable to see planets directly. In this case we must devote still more effort to trying to find planets indirectly, through inductive means.

## PLANET DETECTION VIA INDUCTION

Astronomers are good at attacking old problems in new ways or in using old methods to new advantage. To a fertile mind something as simple as making a better infrared detector to find the infrared emission from extrasolar planets seems hardly inventive. For something really clever try this: Astronomers can find planets by noting their effect on the universe.

One way to do this, as discussed above, is to wait for the planets' inhabitants to visit. Another, likely to prove more fruitful, uses a more conservative approach. We can't be sure that planets produce inhabitants who explore the galaxy, but we *can* be sure that planets produce gravitational forces. This being so, each planet pulls on its parent star (and on everything else) with an amount of gravitational force whose strength varies in proportion to the planet's mass. If we observe the planet's star with extreme care, we may be able to see the result of the gravitational tug from the star's planet or planets (see Figure 16).

But when we say *extreme* care we are not exaggerating. A planet is of course far less massive than the star it orbits. (If it were not, then the star would be said to be orbiting the planet.) Jupiter has a mass only a thousandth of the sun's mass. One might leap to the false conclusion that this means that Jupiter's gravitational force on the sun

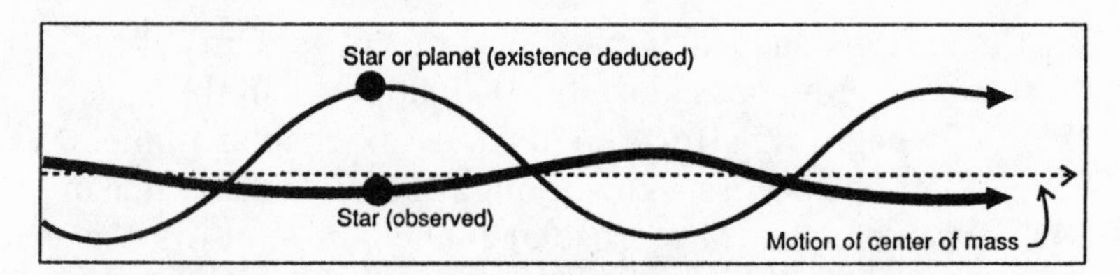


Figure 16: A planet's gravitational pull on its star makes the star move in a tiny orbit as the planet moves in a much larger orbit. The ratio of the sizes of the two orbits equals the ratio of the planet's mass to the star's mass. *Drawing by Crystal Stevenson*

at all times is just a thousandth of the sun's gravitational force on Jupiter. In fact, however, as is expressed in Newton's famed third law ("every action has an equal and opposite reaction"), Jupiter pulls on the sun with exactly the same amount of gravitational force as that with which the sun pulls on Jupiter.

Why, then, don't Jupiter and the sun orbit each other—or approach one another at equal velocities? As we saw in Chapter 2, the difference that the mass makes lies in the way that the objects react to a force upon them. As expressed in Newton's second law ("acceleration equals force divided by mass"), a more massive object will react far less to an outside force than a less massive object will. Hence the key to the motions of a star and its planets is the fact that the planets' relatively tiny masses make them accelerate far more readily than the star to a given amount of force, namely the star's gravitational force on each planet (which equals the planet's gravitational force on the star). It is the planets that move in sweeping orbits around the star, not the star that orbits the planets; but to be precise (and this is crucial for our purposes), both the planets and the star do perform orbits. All the orbits use as their central reference the center of mass of the system, but since the star has most of the mass, this center of mass lies close to the center of the star. Nevertheless the star *does* perform a small orbit around the center of the mass. The challenge is to detect the orbital changes in the star's position, which would signal the existence of a planet.

## THE GRAVITATIONAL WOBBLE

To detect the slight changes in a star's position that a planet causes as it orbits that star is no small feat. If we are trying to find Earth-like planets with this method we had better concede defeat at once, for a planet with the Earth's mass tugs so gently at its parent star that the gravitationally induced changes in the star's location are completely undetectable with our present techniques. But judging our solar system to be—perhaps—typical of planetary systems in the galaxy, we may ask for the likelihood of detecting not an Earth-like planet but a Jupiter-like planet in orbit around nearby stars.

Here the situation appears bad but not hopeless. Jupiter has 318 times the Earth's mass and orbits the sun at five times the Earth–sun distance. The greater distance decreases the force of gravity that Jupiter exerts on the sun by a factor of twenty-five over what the force would be if Jupiter had the Earth's distance. But the greater distance increases the contribution that Jupiter makes to the location of the solar system's center of mass by a factor of five. The net result is that Jupiter's ability to make the sun move from what would otherwise be the sun's location falls to one-fifth (but at least not to one twenty-fifth) of what its ability would be if it had the Earth's distance from our star. This fact combined with Jupiter's mass gives Jupiter about sixty-four times the ability to move the sun that the Earth has.

Suppose then that a Jupiter orbits one of the closer sunlike stars—Alpha Centauri, perhaps, or Procyon—at a distance equal to the sun-Jupiter distance. What are our chances of finding this other Jupiter by studying its parent star's movements?

In order to answer this question thoroughly we must recognize that the (hypothetical) planet's gravitational pull on its star can potentially be detected in two different but complementary ways. One way is to measure changes in the star's position over time as its planet orbits around it.

The other method is to detect changes in the star's velocity, measured along our line of sight (toward us or away from us), likewise over the course of the planet's orbit.

When considering the first, more obvious method—searching for possible changes in a star's position—an important fact must be taken into account: Any such changes caused by a planet's gravitation will appear superimposed on the already significant motion of the star. In other words, we must look not for a change where none should appear, but rather for changes in position superimposed upon other changes. This makes our task more difficult.

## THE PARALLAX EFFECT

Any star's observed position on the sky will change because of two fundamental effects. First, we observe the cosmos from a moving platform, one that circles our own star once per year. This annual motion around the sun gives us a changing view of the sky, since the Earth's location in January and the Earth's location in July lie on opposite sides of the sun. As a result we expect that every star shows the effect of this parallax—the differing appearance of any object when seen from two or more different angles of approach (see Figure 17). Photographers know about parallax, for they sometimes use cameras that have one lens for viewing and another for photographing—which would leave close-up scenes off-center if parallax were ignored.

Astronomers too know plenty about parallax and have learned to use it to their advantage. The amount by which a distant object appears to change its position—the size of its parallax shift—as the Earth orbits the sun depends on the object's distance from Earth. More distant objects show lesser parallax shifts than closer objects since the Earth's back-and-forth changes in position are smaller in comparison with the distance to the object. Astronomers have taken many photographs of stars in the Milky Way

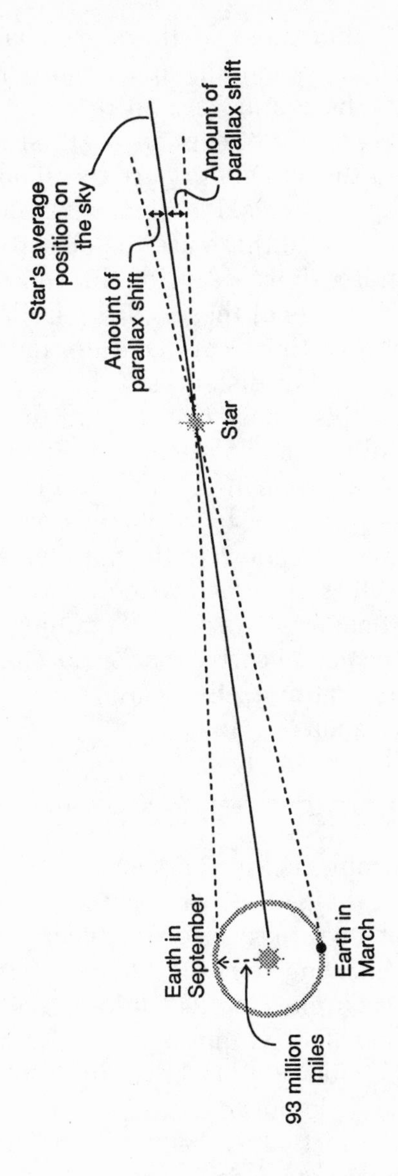


Figure 17: The parallax effect that arises from the Earth's motion in orbit makes the closer stars appear to shift their positions during the course of a year with respect to the backdrop of much more distant stars. *Drawing by Crystal Stevenson*

and have noted that most of the stars that appear on photographs taken six months apart show no detectable parallax shifts. These stars are so far from us that the astronomically modest changes in position arising from our orbit around the sun simply are too small to register on the photographic plates. But some stars do show measurable parallax shifts; these are the closer stars—those less than a hundred light-years of the solar system. By measuring the amounts of these stars' parallax shifts, we make what are still the most accurate and most direct determinations of stellar distances.

Of course, it is just these closer stars that we seek to search for possible planets. Hence when we make our search, taking photograph after photograph in the hopes of finding changes produced by a planet's gravitation, our first task must be correction for the parallax effect, which will change the stars' positions without any planets being present. To astronomers this presents nothing startling as a concept, but the actual correction for parallax inevitably introduces errors that make the search for tiny changes in position more difficult.

## THE PROPER MOTIONS OF STARS

But another problem exists. Every star (and its planets if it has any) in the Milky Way orbits the center of the galaxy. For most stars these orbits are nearly circular. The sun and its neighboring stars all move in nearly the same direction at speeds close to 240 kilometers per second, with the exception of a few interlopers called high-velocity stars. The high-velocity stars have different orbits than the majority: They do *not* move along nearly circular orbits in the same direction as the other stars and therefore have large velocities *as seen from one of the majority of the stars.* The term high-velocity thus refers not to the stars' speeds around the galactic center—which may be no greater than those of the sun or similar stars—but to the stars' speeds *with respect to stars like the sun.*

If all the stars in the solar neighborhood moved in perfectly circular orbits and in the same direction, their motions would produce no additional changes in the apparent positions of nearby stars. Like the Earth's turning surface, which does not make us all lose sight of one another, a perfectly matched set of orbits would leave us with unchanging *relative* positions, like the fixed horses on a merry-go-round. But in fact no star has a perfectly circular orbit, and the deviations from perfect circularity vary from star to star. Hence every star, not only the high-velocity stars, has a proper motion—a motion of its own (for a nearly lost sense of the word *proper* means "one's own") that does not match that of the stars nearby.

In more precise astronomical terminology, proper motion refers to the apparent motion of a star on the sky once we correct for the parallax effect discussed above (see Figure 18). If a star's orbit were perfectly matched to the sun's it would have zero proper motion, but since this could be true only by extreme coincidence, all stars have some proper motion. The actual amount of the observed proper motion depends on two factors: the actual difference between the star's orbit and the sun's and the distance between the star and the solar system. At greater distances a given difference in orbits makes a smaller visual impact; that is, the star's proper motion decreases (for a given orbital difference) in proportion to the star's distance.

The proper motion method has promise, but it takes time. Since Jupiter takes twelve years to orbit the sun, to find a Jupiter-like planet at a Jupiter–sun distance from its star would require a significant fraction of twelve years of observation. Better still would be several orbital periods of observation to be sure that we did not observe some sort of glitch, either because something happened to the star or (more likely) because we made a systematic error in our observations. In short, if you want to find planets by studying stars' proper motions, you had better lay long-term plans.

These plans have indeed been laid, and in important

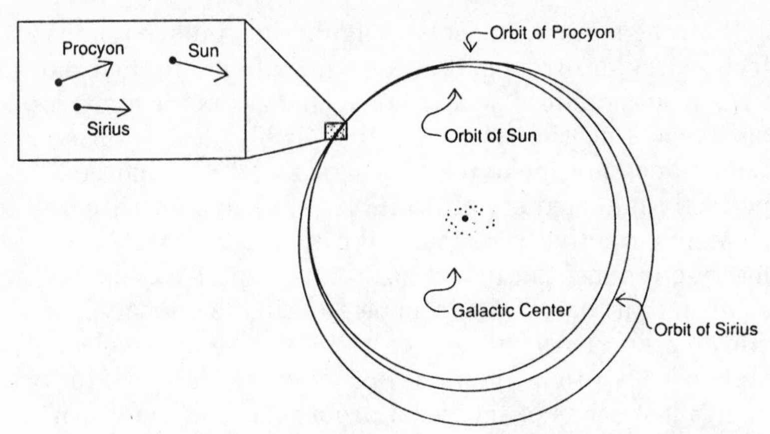


Figure 18: The proper motion of nearby stars such as Sirius and Procyon arises from the relatively small differences between the star's orbit around the center of the Milky Way and the sun's. To detect this proper motion, astronomers must first allow for the parallax effect. *Drawing by Crystal Stevenson*

ways have been executed. The results are tantalizing but inconclusive: It is fair to say that the proper motion observations to date have not yet convinced astronomers that any planetary companions—any objects with Jupiter-like masses—have been found around other stars. But another method, one that also relies on a planet's influence on its parent star, seems to offer brighter prospects.

## PLANET-INDUCED CHANGES IN VELOCITY

The second basic effect that a planet causes on its parent star consists of changes in the star's velocity. As a planet pulls on its star, producing the changes in position discussed above, it also changes the star's velocity, since it tugs the star toward the planet. (More precisely, the planet and star each orbit their center of mass, which lies far closer to the star's center than to the planet's center.) This small orbit by the star produces continuous changes in the star's velocity.

In theory, then, we can detect planets around other stars

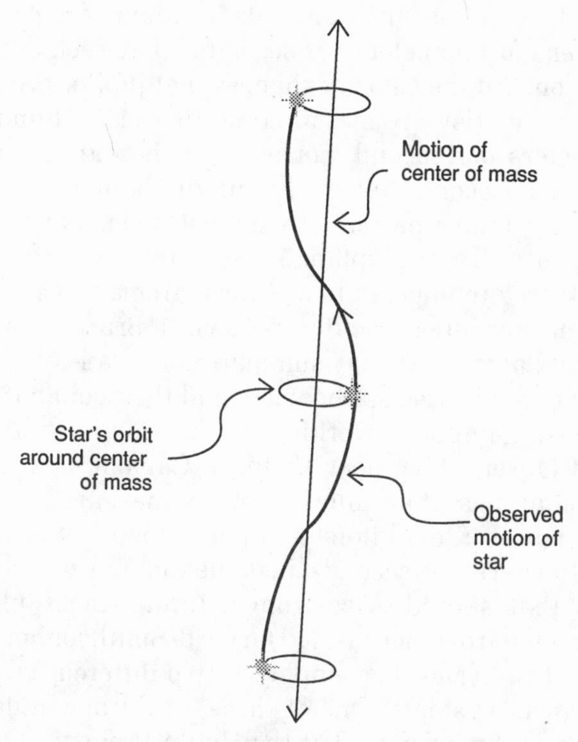


Figure 19: As a planet orbits its star and both orbit the Milky Way, the planet's gravity gives the star a small additional velocity. This velocity will be one of approach when the planet lies between ourselves and the star and one of recession when the planet lies on the far side of the star. *Drawing by Crystal Stevenson*

by carefully measuring the star's velocity through the Doppler effect, which reveals a star's velocity toward us or away from us. As a planet passes (roughly) between ourselves and its star, it pulls the star toward us, and as it passes on the opposite side of its star, it pulls the star away from us (see Figure 19). All we need to do—in theory—is keep measuring the star's velocity, and we will see changes, first of approach and then of recession, induced by the planet's gravitational force upon the star.

The theory is absolutely correct (so far as we can tell), but the changes are unsettlingly small. Astronomers habitually employ the Doppler shift to measure the velocities of stars that range into the hundreds of kilometers per second, or tens of kilometers per second if they seek detailed information. But the velocity changes that Jupiter produces in the sun's motion are measured not in tens or hundreds of kilometers per second, nor even in the individual kilometers per second, but in the hundreds of *meters* per second—less than 1 percent of what astronomers typically can measure. To find planets with the velocity shift method therefore appears to call for extreme measures.

Extreme measures are not unknown to Bruce Campbell and David Latham. We may summarize a decade of careful effort by each of these astronomers and their collaborators as follows. Campbell, working at the Dominion Astrophysical Observatory near Victoria, Canada, has developed techniques that allow him to measure velocity changes not of a few kilometers per second but of a few hundred meters per second—just the amount of velocity changes that should arise from a Jupiter-mass planet. Latham, an astronomer at the Harvard-Smithsonian Center for Astrophysics, has employed two different velocity measurement systems. One of these, of European design and engineering, was used at the Haute-Provence Observatory in southern France and at the European Southern Observatory in Chile; the other, called the digital speedometer, has been installed by the Center for Astrophysics on the sixty-inch telescope at Oak Ridge, Massachusetts, and at the telescope of the Whipple Observatory in southern Arizona.

The independent observing programs guided by Campbell and Latham have yielded tantalizing results. Campbell has found a number of intriguing candidates, stars that seem to have a companion with a mass similar to Jupiter's, but the data still have too much uncertainty to allow a firm conclusion to be drawn. The good news here is that as time passes Campbell will accumulate more data—a longer baseline in time—that will allow more definitive pronouncements to be made.

Latham, on the other hand, has found one fine companion—but one that may or may not be a planet. Of the several dozen stars that Latham has observed repeatedly during the past ten years, he finds a clear signal of a low-mass companion around one called HD114762 (HD stands for the Henry Draper catalog of stars, a reference work familiar to astronomers). HD114762 closely resembles the sun and has a distance of about a hundred light-years from the solar system. This makes HD114762 one of the few hundred thousand closest neighbors of the sun—hardly next door but rather close in galactic terms. A sphere centered on the sun with a radius of one hundred light-years includes only one-millionth of the total volume of our galaxy—and only one-millionth of the total number of stars in the Milky Way. Hence HD114762 is closer to us than 99.9999 percent of all the stars in our galaxy.

Latham and his colleagues made sixty-three observations of HD114762 during a seven-year span and found that the velocity of the star along our line of sight varied by plus or minus 550 meters per second from its average value (see Figure 20). The variation repeated cyclically, with a period of eighty-four days—almost the same amount of time that Mercury takes to orbit the sun. This implies that HD114762 has a companion that orbits relatively close to the star—which helps make the velocity perturbations larger, since the proximity increases the gravitational force that the planet exerts on HD114762. Using Newton's laws of motion, Latham's group can also calculate the mass of the companion, and this mass turns out to be at least eleven times the mass of Jupiter.

Why is the companion's mass available only as a lowest possible mass? We don't know the orientation of the companion's orbit to our line of sight. If the plane of the orbit coincides with our line of sight, then we see all of the velocity perturbation that the companion produces on the star. But if the plane of the orbit is tilted, then some of the velocity caused by the companion is "wasted" so far as our ability to detect it goes. That is, we do not see the full velocity perturbation as a change in the star's velocity

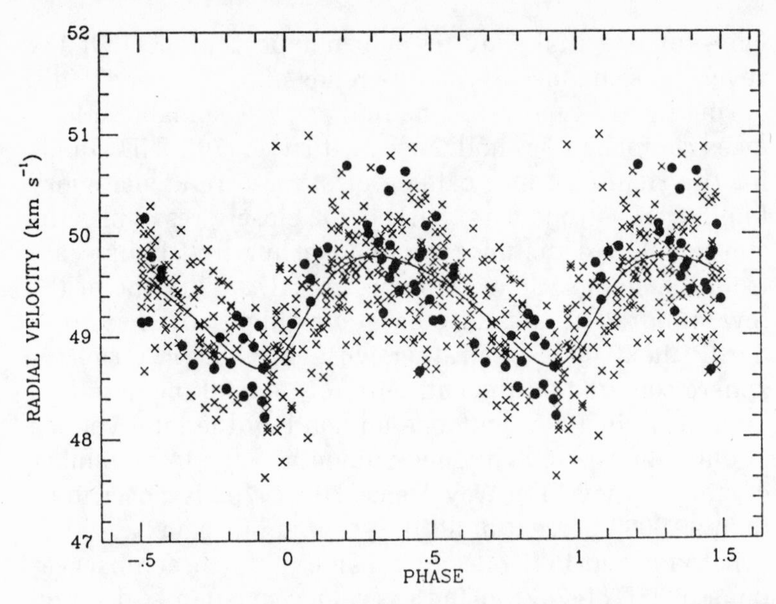


Figure 20: David Latham and his colleagues have measured small, repetitive changes in the velocity of the star HD114762 that indicate a companion with at least eleven times Jupiter's mass in orbit around the star. The line represents the best fit to the observational data points, which show a wide scatter that arises from the difficulty of making accurate velocity measurements. *Courtesy of David Latham*

toward us or away from us because some of the velocity perturbation is always directed across our line of sight (see Figure 21). If some of the velocity perturbation goes undetected, then we might have a companion with larger mass—which produces a larger velocity perturbation—but see only a fraction of the total perturbation produced. This means that we underestimate the companion's mass whenever the orbit's plane does not coincide with our line of sight.

As Copernicus taught us, we can hardly expect that the cosmos happens to be oriented just as we would like it, and we must anticipate that a random orbit plane does *not* coincide with our line of sight. A probability analysis reveals that on the average the orbit plane of a star and its companion will be tilted at about thirty degrees away from

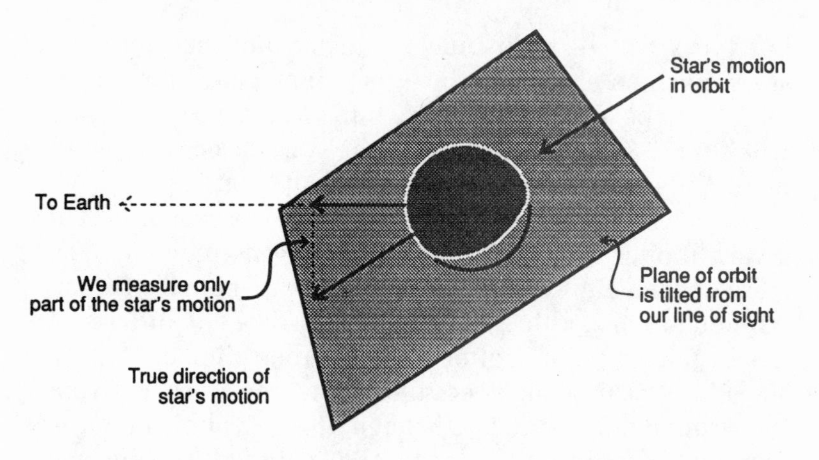


Figure 21: When we measure the change in velocity of a star, we probably measure less than the full change, because in most cases our line of sight does not lie directly in the plane of the orbit of the star and its companion. If the orbit plane is highly inclined to our line of sight, most of the actual velocity will go undetected. *Drawing by Crystal Stevenson*

the line of sight. This implies that on the average the mass of the companion will be about 1.2 times the minimum mass—the mass that we deduce by assuming that the orbit plane coincides with our line of sight.

In other words, on the average when you find a companion whose mass equals at least eleven Jupiter masses, the actual mass will be thirteen Jupiter masses. But HD114762 is not "on the average"; this star represents the best candidate among the several dozen stars followed by Latham and his group. If—and this is a big if—Jupiter-like planets are in fact extremely rare but companions with masses many times Jupiter's are rather common, then it is far more likely that Latham has found a companion with thirty or forty Jupiter masses in an orbit highly tilted to our line of sight than that he has found an eleven- to thirteen-Jupiter-mass companion in a mildly tilted orbit.

This makes a big difference because a companion with thirty or forty Jupiter masses is surely not a planet but a brown dwarf. Brown dwarfs are objects much like stars but too small to begin nuclear fusion and become true stars. Even an eleven- to thirteen-Jupiter-mass companion may well be a brown dwarf, but such a brown dwarf would at least be not *too* unlike the sun's largest planet in mass and composition—a sort of quasiplanet.

But one might object that Jupiter itself is a sort of brown dwarf. Jupiter consists mainly of hydrogen and helium, as stars do, and can be thought of as a companion to the sun with one-thousandth of the sun's mass. The key difference lies in the origin of Jupiter and the other planets. All the sun's larger companions are thought to have formed "from the ground up," that is, through the accretion of large numbers of individual objects that collided, one by one with the object-in-formation. In contrast, the sun—and brown dwarfs—are thought to have formed "from the top down," that is, from a single clump of material that contracted and grew denser within an interstellar cloud of gas and dust. We may think of Jupiter as the total of billions of comets that collided and stuck together, but the sun—and the other stars in the Milky Way—did not form in this way. Instead the stars formed from a single clump, which left trillions of cometlike objects in orbit around the star-in-formation. Many of these comets never accreted anywhere; the sun to this day has a trillion or so comets in orbit around it, far beyond all its planets. But the regions closer to the sun saw an epoch of tremendous cometary accretion four and a half billion years ago, which left behind the Earth, the moon, and the other inner planets; the four gas giant planets; and the smaller clumps of debris that became Pluto, the asteroids, the meteoroids, and the satellites of the outer planets.

In short, in the search for planets brown dwarfs have little to claim though they do suggest improved search methods. Today we remain shy of the first find of a planet beyond the solar system. Let us salute Bruce Campbell and David Latham in their ongoing, painstaking research; let us salute the university centers that fund their efforts; and let us hope that eventually the search allows them to say, "*Here* we have what must be a planet."

# 10

# Double Stars, Black Holes, and Gravity Waves

The Milky Way galaxy contains perhaps fifty or sixty billion places where stars orbit around one another in double-star, triple-star, or higher-multiple systems. So long as the stars in these systems remain in the full flush of life, the systems present few mysteries to astronomers. But when one or more of the stars in such a system grows old and then dies, interesting effects can arise on the stars that are left and in the system as a whole. One of the most intriguing situations occurs when one of the stars explodes and the explosion leaves behind a black hole.

## BLACK HOLES LOCKED IN MORTAL EMBRACE

Black holes are creatures of Einstein's general theory of relativity, described in a few eloquent equations, once believed to be only a theorist's dream and now recognized as significant—potentially the most significant—components of the universe. (The "most significant" description would apply if black holes turn out to form most of the dark matter described in Chapter 13, whose existence astronomers deduce from its gravitational effects. In this case, black holes would constitute most of the mass in the universe.) Throughout the 1970s and 1980s astronomers have grown steadily more confident, though they are not yet entirely convinced, that the Milky Way contains large numbers of black holes, most of them the remnants of former stars. And as this belief in black holes from stellar evolution has strengthened, astronomers have grown more willing to consider what would happen—what may happen rather often—if both members of a binary star system become black holes.

Consider, then, a double-star pair in which both stars are born together, grow old together (because they have roughly the same mass, and more massive stars age more rapidly), and die together in supernova explosions that leave behind black holes with perhaps two to ten times the sun's mass. The expulsion of the stars' outer layers during their red-giant and supernova phases would cause the stars to lose mass and would therefore change the stars' mutual gravitational attraction, which in turn would change the sizes and shapes of the stars' orbits.

But the fundamental fact about those orbits would persist from star birth through star death: The stars would continue to orbit around their common center of mass, the point that lies on the line joining the centers of the two stars, halfway between the stars if their masses are equal, closer to the more massive star if they are unequal (see Figure 22). If, for example, one star had twice the mass of the other, the center of mass of the binary system would lie twice as far from the center of the less massive star as from the center of the more massive star. Wherever the center of mass might fall, each star would perform an orbit around that point in space. A less massive star would move along a larger orbit than its more massive companion star because the center of mass lies closer to the center of the more massive star.

As a general rule, so long as the mass of the stars did not change, neither would their orbits. Astronomers call motions that obey the mathematical rules developed by Isaac Newton from his universal law of gravitation and his three laws of motion Newtonian dynamics. Newtonian dynamics provides solutions to Newton's laws—orbits in space—that can remain the same indefinitely. Thus Newtonian

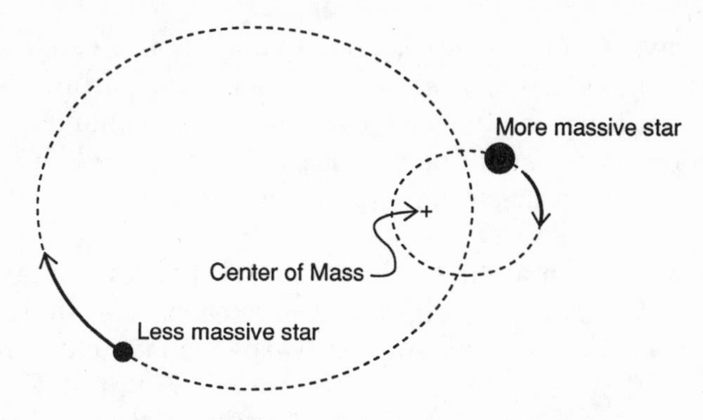


Figure 22: In a binary star system, the center of mass always lies on the line joining the two stars' centers but closer to the more massive star. The ratio of the distances from the center of mass to each star equals the ratio of the masses of the two stars. *Drawing by Crystal Stevenson*

dynamics explains the motions of the planets around the sun, of the moon around the Earth, of the planets' satellites, and of the sun and other stars around the center of the Milky Way—all with near-perfect accuracy and with near-infinite duration. Billions of years may go by, but we expect that the Earth's orbit around the sun and the sun's orbit around the center of our galaxy will barely change. This is the world picture that Newton left us: Celestial objects wheeling along majestic, unchanging trajectories, a universe perduring. As Einstein once wrote in his private notes (given here in a loose translation from the German by the authors):

> Look unto the stars to teach us
> How the master's thoughts can reach us:
> Each one follows Newton's math
> Silently along its path.

But it was Einstein who showed that for the closest stellar pairs, those most tightly locked in gravitational embrace, the Newtonian picture must be abandoned. In its place, according to Einstein's general theory of relativity, we must picture a universe in which gravity bends space, and bends it most strongly in regions closest to the source of gravity—for example, close to a black hole. As a result of the bending of space, when one black hole orbits close to another it can no longer move along one unchanging orbit indefinitely. Instead bent space will cause the black holes eventually to meet one another, merging their existences forever.

Well, the amateur student of black holes may say, and good. If black holes orbit so close to each other that one of them encounters the highly warped space close to the other, they may indeed merge together. But surely this cannot occur often, in view of the fact that for such a merger to occur one black hole must approach the other within two or three black-hole radii. Since stars orbit each other at distances many times the diameter of each star, and since stars have diameters thousands of times larger than black holes, surely a situation in which two black holes come within a few black-hole radii of each other must be extremely rare. A binary star system that evolved into two black holes would show the two separated by something like a hundred thousand times the black-hole radius of either one. This would imply that the two black holes can orbit one another—more precisely, orbit their common center of mass—forever, just as Newton predicted.

But they can't. And they can't because of Einstein. The same theory of general relativity that predicts the existence of black holes also predicts the effect that will eventually make the two black holes merge: gravity waves.

## GRAVITY WAVES

Gravity waves, more formally known as gravitational radiation, are a type of wave entirely unlike the more familiar waves that include light, radio, and all other forms of electromagnetic radiation—gamma rays, x-rays, ultraviolet, infrared, and microwaves. Furthermore, gravity waves are completely undetected: We have yet to verify the first definitive observations of any such radiation. Yet gravity waves are an important component of contemporary astrophysical research; many scientists confidently expect that before this millennium ends we shall have our first gravity wave detection and shall thereby open to our view an entirely new way to look at the universe.

Gravity waves arise from the rapid motion of large amounts of mass in close proximity to one another. Stated more precisely, any motion of two masses with respect to one another, except in certain unlikely, highly symmetric situations, produces gravity waves. However, the larger the masses and the more rapidly they move with respect to one another, the greater the energy carried off by the gravity waves produced by the motion. Since gravity waves—even those that carry enormous amounts of energy—are extremely difficult to detect, from a practical point of view we need consider only those situations that produce the largest gravity waves, those that carry the greatest amounts of energy. These arise from the most rapid motion of the largest masses we can picture in such motion. In other words, the search for gravity waves involves combing the universe for the most violent situations on the largest scales of size and mass.

Consider, for instance, the collapse of a star's core that begins a supernova explosion. In less than a single second the core—which contains several times the mass of the sun—shrinks in size from a diameter of several thousand miles to a diameter of only a few miles. That is, the collapse moves several solar masses of material by several thousand miles in less than a second. According to Einstein's theory, if this collapse proceeds in perfect symmetry, with each part of the core falling inward at precisely the same rate, then no gravity waves will be produced. But if the collapse is nonsymmetric—if different parts of the core fall in at somewhat different rates, perhaps because the core is rotating—then gravity waves will be produced. In that case, the collapse involves large amounts of mass (the different regions of the collapsing core) moving with respect to one another in a way that keeps the waves from canceling one another. Einstein's theory tells astrophysicists how to calculate the amount of energy released in gravity waves during the collapse, and that the amount is rather high—about equal to the amount of energy that the sun radiates during ten billion years.

But there's a catch: Gravity waves are not easily detected. We know that stars explode at a rate of about one per century in a galaxy like our Milky Way, and we believe (at least those of us who believe Einstein do) that each collapse releases as much energy in less than a second as the average star releases in ten billion years. That is, during the second containing the collapse the star radiates as gravity waves about a billion billion billion ($10^{27}$) times more energy than a sunlike star releases during that second in the form of visible light.

## INDIRECT DETECTION OF GRAVITY WAVES

Even though astronomers have yet to detect gravity waves directly, they can already point to one location in the Milky Way where they are certain that they have observed the *effects* of gravity waves. The system known to emit gravity waves is a binary pulsar—that is, a pulsar (described in Chapter 8) in orbit with another collapsed star. This particular system, in the constellation Sagitta, goes by the astronomical name of PSR 1913+16 (PSR for pulsar, the numbers for its astronomical coordinates). In PSR 1913+16, two objects orbit one another at close proximity and with great rapidity. Astronomers have deduced these facts through precise timing of the arrival of the pulses from the pulsar. The intervals between pulses grow longer and then shorter in a repetitive pattern whose characteristic shape indicates—to astronomers—only one possible origin: The pulsar is moving in orbit. When the pulsar moves away from us the interval between observed pulses increases, and when it approaches us that interval decreases. Using this principle, astronomers have deduced the key facts about the orbit in which the pulsar moves: The orbital period equals only 7.75 hours, the orbit has a highly elongated shape, and the average distance between the pulsar and its companion is only forty-three thousand kilometers—barely one-sixth of the distance from the Earth to the moon.

In PSR 1913+16, a pulsar moves in orbit with a companion that is either a black hole or a neutron star that does not produce observable pulses. Here we have the ideal situation to produce gravity waves: the rapid motion, in noncircular orbits and at close proximity, of two objects with significant masses comparable to the sun's mass. Einstein's relativity theory predicts that the two objects must produce gravity waves that will continuously carry away energy from the binary system. The loss of energy will make the two objects move slightly closer to one another and orbit even more rapidly around their common center of mass.

Precisely this effect has been observed and with precisely the amount predicted by Einstein's theory. Ever since this binary system was discovered in 1974, pulsar experts have timed the pulses' arrival and thus have continued to monitor the pulsar's orbital motion. The astronomers have found that the orbit is shrinking exactly as theory predicts. The emission of gravity waves should eventually shrink the orbit to the point that the two objects merge. Meanwhile, in an indirect sense PSR 1913+16 has furnished the first detection of gravity waves—a detection through their effects and not by direct observation. In order to "see" gravity waves pure and simple we must understand more about what they are—and what they do to a possible detector system.

## WHAT ARE GRAVITY WAVES?

Einstein's general relativity theory tells us that gravity waves are a bending of the fabric of space-time itself, ripples that propagate outward from the disturbance (the rapid motion of matter) through space-time like the ripples on the surface of a pond. You can't see them any more than you can see space-time, but you can sense the passage of the gravity waves—if you're careful. A gravity wave distorts an object as it passes by, first squeezing it in one direction while elongating it in a perpendicular direction, then elongating it in the first direction while squeezing it in the second (see Figure 23). Hence if astrophysicists can measure the shape of an object with high precision they can tell when a gravity wave is passing through it, because they will see it grow distended in an alternating pattern of extension and contraction along two perpendicular axes.

A gravity wave detector thus consists of a set of objects, carefully observed, that will (if a gravity wave passes by) change size and shape in the way we have described. If, for example, astronomers seek to observe the gravity waves produced by a supernova in the Milky Way, they must create such a detector and wait (on the average) about a century. Funding agencies won't buy such a proposal, so those who take gravity wave detection seriously do not plan on such a wait. Instead they have extended their horizon of expectation far beyond the Milky Way. They note that the Virgo Cluster, the nearest large cluster of galaxies, contains about a thousand members, each of which may produce a supernova every century or so. If, therefore, they can construct a gravity wave detector capable of noting the gravity waves from a supernova in the Virgo Cluster, they should have less than a year to wait (for a thousand host galaxies are each capable of providing one supernova per hundred years) before the alarm rings and fame becomes assured. And so to gravity wave sensualists the question becomes: What does it take to make a detector that can find gravity waves from supernova explosions in the Virgo Cluster?

The answer is time, money, and brains—especially the latter. The problem seems formidable. A supernova in the Virgo Cluster will—we can calculate—produce gravity

Photo 10: The sun's surface seethes at a temperature of 6,000 degrees Celsius. Much hotter prominences rise above the surface, which shows complex motions that result from the heat rising from below. *Photograph courtesy of Big Bear Solar Observatory*

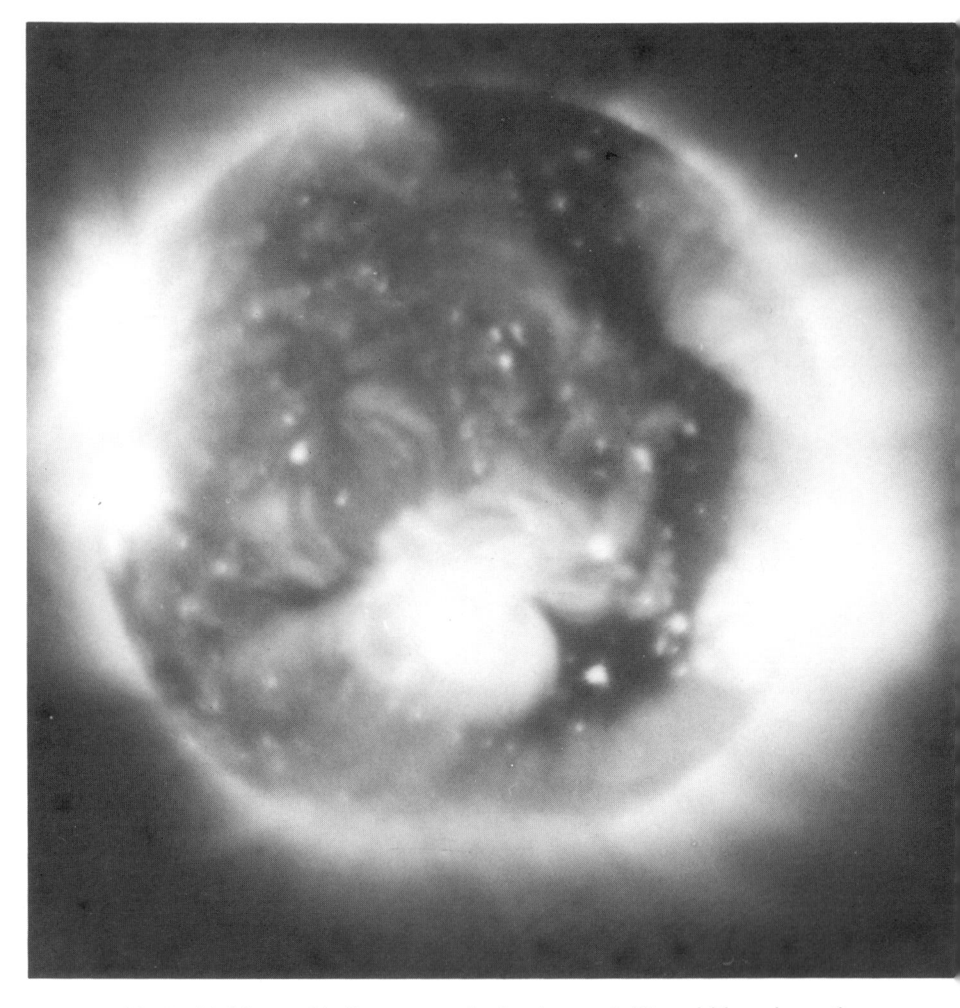

Photo 11: Mapped in its x-ray emission by a satellite orbiting above the atmosphere, the sun looks quite different from the familiar visible object shown in Photo 10. *Photograph courtesy of American Science & Engineering*

Photo 12: The Very Large Array consists of twenty-seven radio dishes that spread over the plains of central New Mexico for twenty miles, forming a radio interferometer that can see as clearly as a single dish twenty miles across. *Photograph courtesy of the Very Large Array*

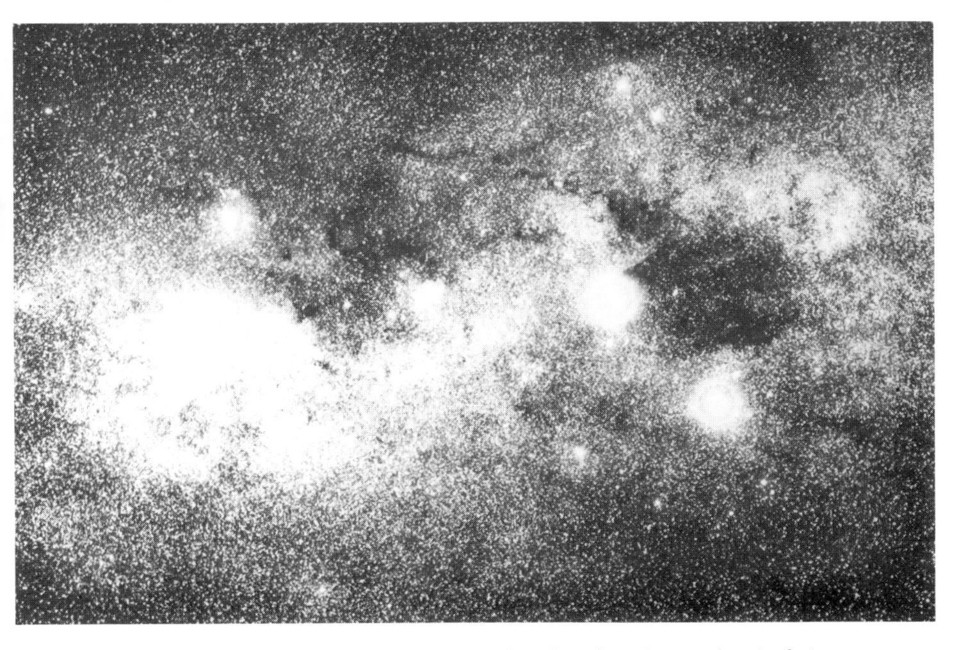

Photo 13: Foreground gas and dust plus the glare from a host of stars obscure our view of the galactic center in visible light. *Yerkes Observatory*

Photo 14: During the 1930s, Karl Jansky, working quite independently of professional astronomers, discovered radio static from the direction of the galactic center. *Photograph courtesy of the Bell Laboratories*

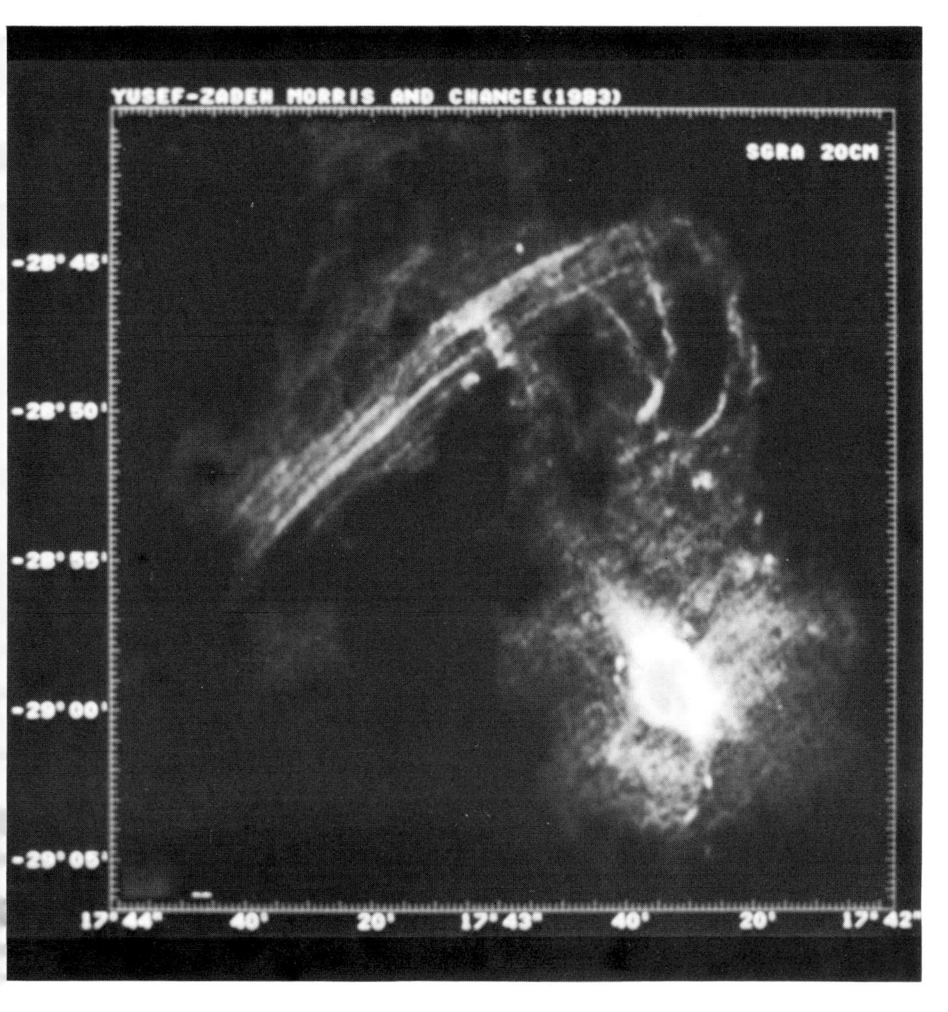


Photo 15: A radio map of the galactic center reveals complex streams of gas. *Courtesy of the National Radio Astronomy Observatory*

Photo 16: The radio emission from a classic double radio source (in this case the galaxy NGC 5128) exhibits jets in opposite directions. This map shows only the innermost part of the radio-emitting region. *Courtesy of the National Radio Astronomy Observatory*

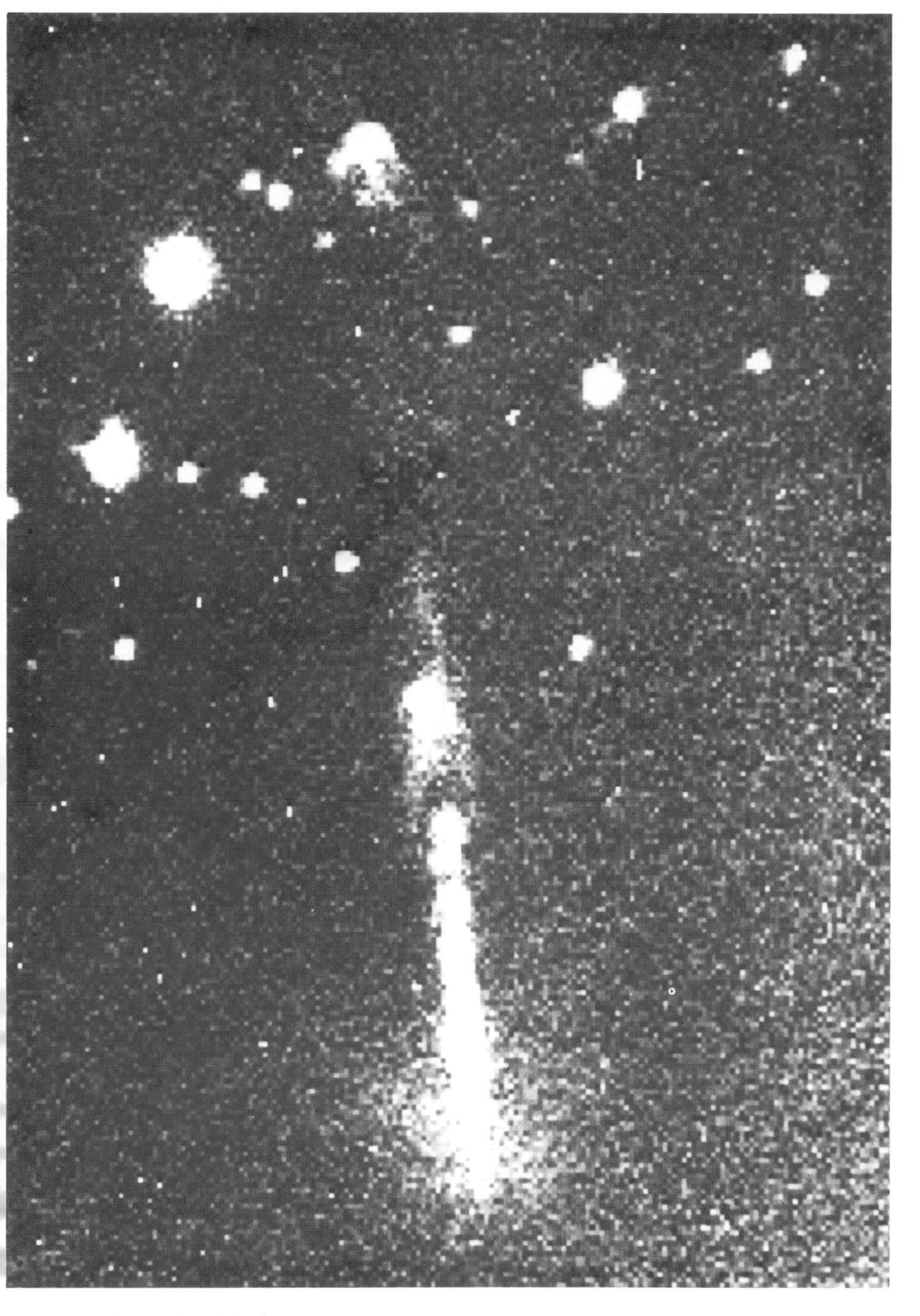

Photo 17: A T Tauri star, a young solar-type star, is shown here (bottom) with a jet of gas (upward) that points to nearby Herbig-Haro objects (top), which are presumably parts of the cloud from which the star has recently formed. *Photograph courtesy of Bo Reipurth, European Southern Observatory*

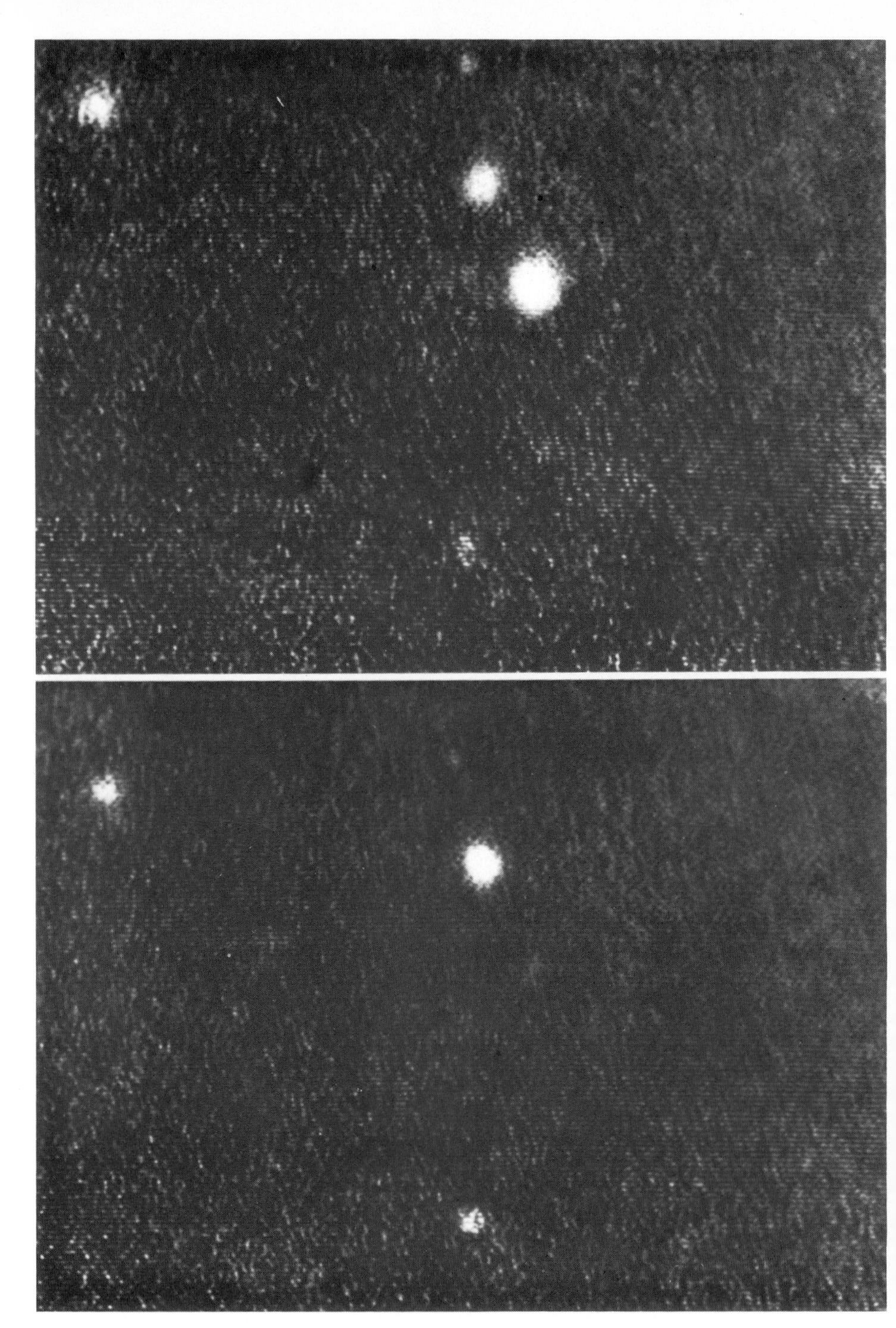

Photo 18: At the center of the supernova remnant called the Crab Nebula, a pulsar flashes on and off thirty-three times per second. These photographs capture the pulsar in visible light during the "on" portion (top) and "off" portion (bottom) of its pulsation cycle. *Lick Observatory*

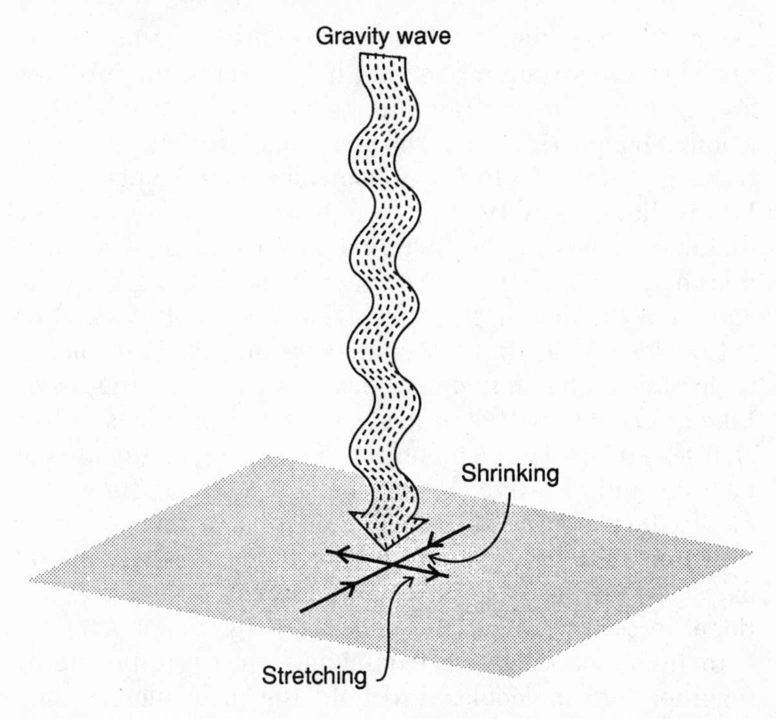


Figure 23: A passing gravity wave will stretch objects in one of the two directions perpendicular to the wave while compressing them in the other direction. Then compression will occur in the first direction and stretching in the other, and so on in alternation. *Drawing by Crystal Stevenson*

waves on Earth that distort the objects in the detector by one part in $10^{21}$—that is, one part in a thousand billion billion. To observe such a tiny distortion amounts to finding a change in size by less than the diameter of a human hair in a system that would reach from the Earth to Alpha Centauri! Or to put this another way, the required gravity wave detector must note changes in distance by less than the diameter of an atomic nucleus even if the detector is ten miles long.

The reason that gravity waves produce such tiny distortions lies in the relative weakness of gravity. The forces that particles exert on one another by gravity pale in comparison with other types of forces, such as the electromagnetic forces that keep electrons in orbit around atomic nuclei or the strong forces that hold together the protons and neutrons in an atomic nucleus. At the size level of elementary particles such as electrons, protons, and neutrons, gravitational forces are so weak as to be negligible. What allows gravity to dominate the universe at large distance scales is the fact that gravity acts over large distances (though it grows weaker in proportion to the square of the distance), and gravity always attracts.

Consider what that last statement means. Electromagnetic forces come in two varieties, attractive and repulsive: Like charges (positive or negative) repel, whereas unlike charges attract. Hence an object made of equal amounts of positive and negative electric charge—a planet, for example—will exert *zero* electromagnetic force on an electron or a proton. But because *every* tiny particle in the planet exerts an attractive gravitational force, the total gravitational force from the planet reaches a significant level.

In human bodies, electromagnetic forces bind atoms together into molecules and hold the molecules in long chains. Furthermore, we draw our energy from chemical reactions that are at bottom the result of electromagnetic forces. When a high jumper springs over the bar, for a moment the electromagnetic forces at her disposal—the energy she draws on as she runs and leaps—outbalance the gravitational force that the Earth exerts upon her. Thus despite the fact that the Earth contains about $10^{23}$ (a hundred billion trillion) times more particles than the high jumper, *her* particles temporarily outduel the Earth's!

It is this fundamental weakness of gravity—the fact that on the smallest scale of sizes, gravity is the weakest of all known forces—that makes gravity waves hard to detect. An electromagnetic wave tends to move charged particles much as a gravity wave moves and distorts particles with mass. But because electromagnetic forces are inherently much stronger than gravitational forces, we can easily move the electrons in a radio antenna with an electromagnetic wave from a radio beacon many miles away. With gravity waves we could barely distort a much more sensitive antenna even if we moved million-kilogram masses at nearly the speed of light.

Thus any gravity wave detector capable of success must have tremendous sensitivity. Even so, the detector must be observed patiently and carefully in order to detect a passing gravity wave. To detect gravity waves from supernovae in the Virgo Cluster, the detector must be far more sensitive than anyone believed possible until recently. Such detectors are now in the planning stages. We owe the possibility of their existence to the ideas that have germinated in the fertile imaginations of experimentally oriented gravity-wave physicists—most notably to Ron Drever and Rainer Weiss.

## THE FUTURE OF GRAVITY WAVE DETECTORS

To make a detector on Earth capable of detecting gravity waves from the far reaches of space you need insight and energy. Such qualities appear in the two collaborating research groups at Caltech and the Massachusetts Institute of Technology (MIT), led by Ron Drever and Rainer Weiss, respectively. These two physicists have pioneered new and imaginative ways to detect gravity waves, which—Congress willing—are to be embodied in two detectors scheduled to be built during the mid-1990s. Weiss and Drever each deserve credit for inventing this device (more precisely, for inventing the details that make it work), but in the search for financial support Caltech proved more willing than MIT. As a result the detector's headquarters are in Pasadena, California, and not in Cambridge, Massachusetts.

The project created by Drever and Weiss is called LIGO, short for Laser Interferometer Gravitational Wave Observatory. LIGO's basic idea is rather simple: Not one but two gravity wave-detecting instruments called gravity wave interferometers are built on opposite sides of North America. Separating the two detectors provides an increased ability to determine whether a disturbance is a gravity wave from deep space or a bit of "noise" from somewhere on Earth. Any local disturbance—a large truck or a small earthquake, for instance—represents the sort of "noise" that could fool a single detector into believing that it had found a gravity wave. But two detectors three thousand miles apart can separate noise from actual gravity waves by having a computer continuously check what both detectors register and pay attention only to those events that they both register simultaneously.

So far, so good: The system eliminates most of the "noise" problem; all that remains is to detect the actual gravity waves. This requires a detector about a thousand times more sensitive than the best existing detectors. Of course not one but two of these detectors must be built, and the budget for the total system must (for science-politics reasons) not exceed some $200 million, the cost of a giant telescope.

The most sensitive gravity wave detectors built until now are large aluminum rods, several tons in weight, that have been hung with extreme care and that can be monitored with incredible precision (see Photo 22). But these gravity wave detectors, most experts agree, have not yet proved sufficiently sensitive to detect gravity waves. Today gravity wave physicists can detect changes in the distance between the two ends of such a bar—which may be some ten feet apart—to better than one part in a billion billion ($10^{18}$). This amounts to measuring changes in this distance of less than a million-billionth ($10^{-15}$) of a centimeter! Incredible though this ability may seem, it comes nowhere close to providing the sensitivity needed to detect gravity waves from collapsing stars in the Virgo Cluster.

The first step toward making sufficiently sensitive detectors is to abandon the notion of simply making a longer bar. Because we need to improve the sensitivity of the detector about a thousandfold, we must create bars whose lengths would be measured not in meters but in kilometers. Today we cannot plan any such apparatus, for we cannot forge such a five-kilometer-long bar or monitor its position with anything like the required accuracy.

But we need the *equivalent* of a bar several kilometers long; that is, we must construct *something* that long, and then measure changes in its lengths to a precision of about a million-billionth of a centimeter. Rainer Weiss proposed using not a metal bar but a beam of laser light as the basic measuring unit. But how then to detect tiny changes in the length? The trick that Weiss and Drever found was a way to let the beam of laser light measure its own length.

This is where interference enters the picture. Interference describes the fact that when two beams of light waves—or of any other types of waves—meet each other, they tend either to combine their effects into a larger total or to cancel each other's effect to a total less than that of either wave. If the waves are in phase, so that the peak of one set of waves matches the peak of the other set, then the effects enhance each other (see Figure 24). But if the two wave sets are out of phase, so that the wave crests of one meet the wave troughs of the other, then the waves tend to interfere; that is, they cancel each other. This principle of interference has been known to physicists and used by them in complex ways for more than a century. During the 1880s Albert Michelson used the interference principle to obtain accurate measurements of the speed of light, a feat that made him the first American Nobel prizewinner in physics.

Drever and Weiss saw that they could use the interference principle to accomplish their ends by splitting a beam of laser light into two parts and sending the two parts in two perpendicular directions (see Photo E). They could do this by passing the light through a beam splitter, which reflects only half of the light that strikes it while sending the other half through the mirror in the original direction. The two beams of light are meant to travel many kilometers of distance before they each strike a reflecting mirror that reverses their directions. The apparatus is constructed

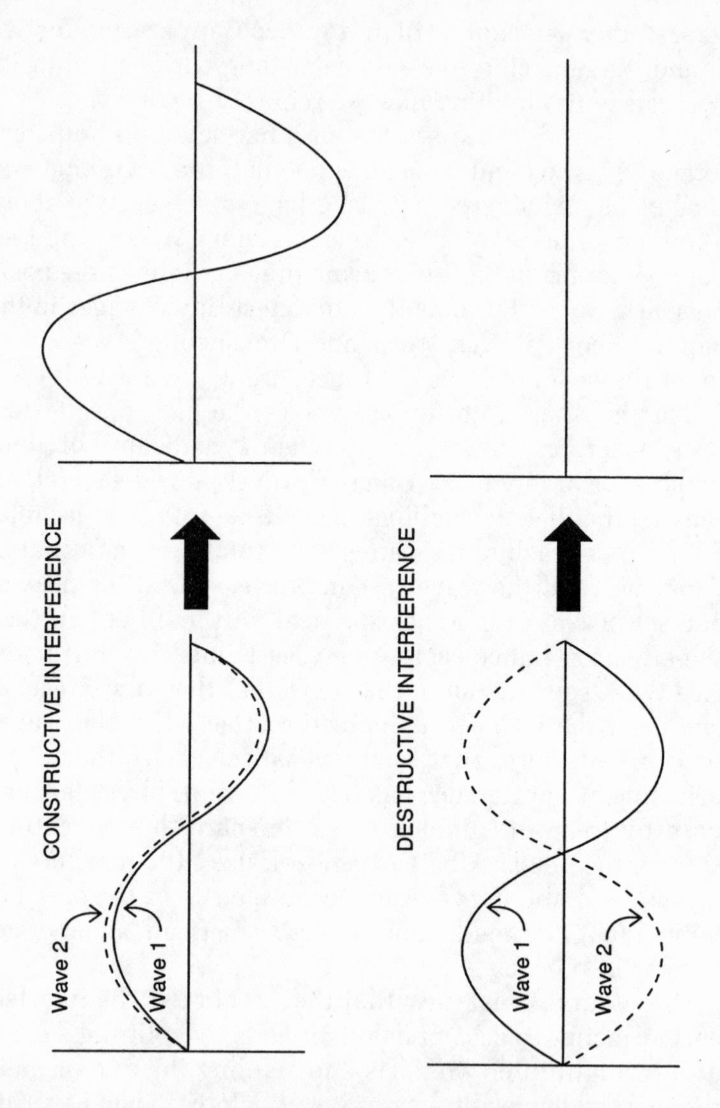


Figure 24: When two waves meet, they can cancel one another in destructive interference if the first wave's crests meet the other wave's troughs, and vice versa. Conversely, if the crests match crests, the waves will add together in constructive interference. When a gravity wave spoils the perfect destructive interference, a spot of light appears where exact cancellation formerly occurred. *Drawing by Crystal Stevenson*

so that if the distances along the two paths are precisely the same, upon their return to the beam splitter the two beams will be exactly out of phase and will *cancel* one another. Then nothing will be visible: the sign that the two beams have traveled the same distance.

But if the distance along one path becomes slightly greater than the distance along the other, the beams will not cancel one another perfectly. Instead, a small spot of light will appear—a sign that something has changed the length of one beam's path in comparison to the other's. How large a change of distance will produce this effect? If one path exceeds the other in length by *one-eighth of the wavelength* of the beams of light, then the constructive interference produces the maximum brightness. This means that the interferometer—the device that splits a light ray into two parts, which travel along different paths before meeting again—is particularly sensitive to distance changes of about one-eighth of a wavelength of the light used in the interferometer. If red light is used, as in many lasers, the wavelength is about six hundred-thousandths of a centimeter, so one-eighth of a wavelength is a bit less than one hundred-thousandth of a centimeter.

This might seem good enough for Drever's and Weiss's purposes, but it is not. With an interferometer of the type described above, Drever and Weiss could not even surpass the performance of the best existing gravity wave detectors. But here Drever and Weiss showed true canniness. With an interferometer that includes two paths, each a few kilometers long, down which the light passes back and forth, the light need not travel back and forth only once. Instead Drever and Weiss have arranged, by making a well-crafted optical system, for the two beams of light to recombine only after traveling down each arm and back again many thousands of times, each time reflecting from a slightly different spot on the mirrors.

Each extra down-and-back journey adds several more kilometers to the total distance that the light travels. This total is exactly what is measured when the two light beams finally recombine. Because the distance increases, the same sensitivity in distance measurement as before lowers the *ratio* of any distance change measured to the total distance traveled by the light. In other words, a longer path length increases the sensitivity of the detector since a change in position that would have gone undetected with a single down-and-back trip may well show up if the light makes a thousand journeys down and back each arm of the interferometer, each time covering a tiny additional distance because a gravity wave is passing through the apparatus.

Drever, Weiss, and Rochus Vogt, who oversees the LIGO project, now propose that LIGO will consist of two systems, each with arms four kilometers long, one in California and one in the eastern United States. Furthermore, the two gravity wave detectors are to embody *four* separate interferometers in each part of the transcontinental LIGO system. The desire for four interferometers arises from the desire to detect gravitational waves of different sorts, as described below, and does not add a tremendous amount to LIGO's cost. The four interferometers would operate side by side in the 1.2-meter-diameter pipes that are to be buried underground and pumped free of air, which would slightly distort the light beams as they make about *one hundred thousand* trips down and back each arm. Traveling at the speed of light, the beams would require a grand total of nearly three seconds to perform their journeys! With this system in place, we should—according to our best current calculations—have a good chance of detecting gravity waves from collapsing stars fifty million light-years away.

## THE SIGNIFICANCE OF GRAVITY WAVE DETECTION

If Drever and Weiss succeed in obtaining some $200 million of federal funding and then report to us before the millennium arrives that they have indeed found gravity waves, they will not have produced a scientific revolution. Yet the detection of gravity waves will set various hearts aflutter, and—still more important for investing money wisely on experiments—if Drever and Weiss *fail* to detect gravity waves, this too will be important.

Gravity waves, as we have noted, are required to exist by Einstein's general theory of relativity. That theory has passed a number of tests, and scientists for that reason have grown fond of it. The general theory of relativity says that we can regard gravity as warping the fabric of space-time, so if we move a source of gravitational force—a massive object—we make ripples in space-time that propagate outward. The theory also tells how strong these ripples will be for a given amount of mass moved at a given speed, and how strongly (more precisely, how weakly) those ripples will interact with any mass that they encounter. Hence when an astrophysicist tells us that collapsing stars, which we believe we now understand fairly well, will produce a certain amount of gravity waves, scientists tend to believe it. If this is wrong, it is possible that Einstein's general relativity theory is wrong, or at least in need of substantial modification. Those are high stakes to consider in the world of experimental physics.

But if Einstein was right—and he has a fine track record so far—we shall still learn a good deal from the detection of gravity waves, beyond any additional support such detection can provide to Einstein's theory. Most important, we shall have a completely different way to observe the universe. Gravity waves are not light waves or any other type of electromagnetic waves; they are in essence an entirely separate phenomenon. If you can detect the gravity waves that a violent event produces, you have a chance to see things about it that you simply could not—not even in theory—without a gravity wave detector. For example, we could learn a great deal about what happens when two black holes collide in a distant galaxy, or just how a star's interior collapses to start a supernova explosion, if we could detect and analyze the gravity waves produced by such events.

The history of astronomy has shown over and over that new techniques to study the universe invariably lead to new discoveries—indeed, they typically produce discoveries of entire new classes of objects and occurrences previously unknown and unsuspected. If gravity wave astronomy soon becomes a reality, we may see this assertion verified once again. Gravity waves offer an entirely new way to see the universe. Almost certainly this new method, once it becomes available, will reveal aspects of the universe only dimly imagined by us now.

# 11
# Failed Stars

Among the hottest topics in astronomical research today is the hunt for brown dwarfs, objects that lie at the boundary between stars and planets. In essence, a brown dwarf is a study in the difference between possibility and actuality: It may be thought of as forever on the way to becoming a star yet never quite capable of doing so. Brown dwarfs may turn out to be the most common type of object in the Milky Way and, by implication, in the universe. Or they may turn out to be relatively scarce. Possibly they do not exist at all—but they ought to, and so astronomers are deep in the search for these incredibly dim, not-quite-stars.

## WHAT IS A BROWN DWARF?

Astronomers reserve the name *star* for an object within which nuclear fusion is occurring now or has occurred in the past. That prototypical star, our sun, produces light and heat because, in its innermost core, nuclear fusion continually turns hydrogen nuclei (protons) into helium nuclei, many mountains' worth per second. Each proton, like every other particle with mass, has energy of mass in an amount given by Einstein's well-known equation, $E$ (energy of mass) equals $m$ (mass) times $c^2$ (the speed of light squared). Nuclear fusion simply converts some energy of mass—energy contained in the hydrogen nuclei that meld together—into kinetic energy (energy of motion). That is the essence of a star; all else is the gaseous wrapping of the nuclear-fusing core. However, we must still seek to explain how nuclear fusion ever began in the hearts of stars, that is, how a star ever acquired a nuclear-fusing core.

The secret of both stars and brown dwarfs is gravity, the most far-reaching of forces, always attractive, capable of organizing the universe over billions of years' time. Stars formed from clumps of gas and dust within interstellar dust clouds. They did so because part of the cloud (the future star and its environs) became a region denser than an average part of the cloud. The greater density induced contraction through gravity, and the contraction produced a higher density. This in turn induced more rapid contraction, because the clump's higher density—by definition, more matter within a given amount of volume—caused each part of the clump to attract all the other parts more forcefully, since the parts were now closer together. Gravitational forces among these parts thus made the contraction proceed more and more rapidly and squeezed the clump to a progressively higher density, most notably at its center.

The increase in density produced an increase in temperature, because matter that is compressed tends to grow hotter, much as the air forced into a tire at high pressure does. The higher temperature means that the particles within the clump have higher velocities as they dash to and fro in random motion. Once the temperature rises to about ten million degrees on the absolute temperature scale, the particles' velocities become large enough that some of the particles would fuse together upon collision. So it was with our sun five billion years ago, or with any other object that squeezes itself to the point that its central temperature rises to ten million degrees and produces a star.

But just as massive objects produce tremendous gravitational squeezing and therefore become stars, smaller objects produce more modest squeezing as their constituent parts interact through gravitational attraction. For this reason, more massive stars have higher central temperatures than less massive stars and produce far greater rates of nuclear fusion, which in turn leads to much larger luminosities—rates of energy output from nuclear fusion. Turning to the lower end of the mass scale, we can see that an object with less mass than the sun will have a lower central temperature than our sun's, somewhere around eight million degrees instead of the sun's fifteen million. This is the case for a star with about half the sun's mass. In such a star, nuclear fusion does occur, but it proceeds at a glacial pace in comparison to a sunlike star. A star whose core temperature reaches only *half* of the sun's fifteen million degrees will have a luminosity only *one-thousandth* of the sun's. Inside a star with still less mass—say 10 percent of the sun's mass—particles barely collide with sufficient energy for nuclear fusion to occur. As a result the star's energy output amounts to a pale imitation of a hotter star such as our sun.

This describes stars with sufficient mass to squeeze themselves to the point that nuclear fusion does begin. Objects with still lower masses will have central temperatures that fall short of the value at which *any* nuclear fusion will occur. And what is that mass? Calculations—as precise as modern astrophysics can make them—lead to the conclusion that *the minimum mass of any star is about 7 percent of the sun's mass.* If a star has this minimum mass—about seventy times the mass of the planet Jupiter—then nuclear fusion will occur (just barely) in the star's core. If an object has a mass less than 7 percent of the sun's, it will never begin nuclear fusion. Such a substellar (less mass than a star) object is a brown dwarf.

Unlike a star, a brown dwarf goes on shrinking. Inside an actual star, the onset of nuclear fusion—which occurs once the star has produced a temperature sufficiently high that some of the nuclei fuse together when they collide—stops the further contraction of the star. Instead, for billions of years (in a typical star) the kinetic energy released through nuclear fusion opposes the star's tendency to contract under its self-gravitation. As a result the star maintains a constant size so long as it has hydrogen nuclei ready to fuse together.

In a brown dwarf, by contrast, nuclear fusion *never* begins. Hence the brown dwarf continues to contract. If nature were simpler than it actually is, the contraction might continue indefinitely until the brown dwarf became a black hole. In real life, so far as we can calculate, the brown dwarf contracts until it becomes so dense that the exclusion principle supports it against its own gravity and halts any further contraction. We can consider the exclusion principle and its effects on matter when we deal with white dwarfs, the cores of former stars, which themselves are supported against contraction by the same principle. For the time being the important fact is that brown dwarfs exist *as* brown dwarfs—objects contracting and growing warmer in their cores—for an astronomically brief time, a few hundred million years or so. After that they essentially sit around peacefully, soon cooling to low temperatures, dissipating the heat generated by simple contraction. So far as nuclear fusion goes, they are failures.

That is the theory. If it is correct, any object with a mass less than 7 percent of the sun's should be not a star but a brown dwarf. Technically, the sun's four giant planets, Jupiter most of all, might be called brown dwarfs: They too are slowly contracting under their own gravitational forces, heating their interiors as a result and radiating this heat into space. Jupiter-like objects would be interesting to detect whether we call them giant planets or brown dwarfs. But still more interesting—and of more potential significance to the universe—would be what we might call "true" brown dwarfs, objects with not one-tenth of a percent of the sun's mass (like Jupiter) but with a few percent of the sun's mass. What fascinates astronomers in their search is that *brown dwarfs might be so abundant that they comprise much, if not most, of the mass in the Milky Way galaxy.*

Why would brown dwarfs be so abundant? Compare the number of stars with the number of planets in the solar system: one to nine! It may prove far easier for nature to make small objects than large ones from a clump of matter such as the one that formed our solar system. When astronomers compare the relative numbers of stars with different masses they find that the low-mass stars far outnumber those with middle masses, such as our sun, which in turn far outnumber the high-mass stars. Although we have no data on brown dwarfs—since we can't see them—an extrapolation of the data on true stars suggests that brown dwarfs may be far more numerous than stars of any type in the Milky Way.

Though astronomers have barely found *any* brown dwarfs, they know that the nature of brown dwarfs makes them extremely difficult to detect, so we may have overlooked them for years. Astronomers may yet find that brown dwarfs are not just one inhabitant of the galactic zoo, but instead represent a major component—perhaps *the* major component—of our galaxy, and by implication of the entire universe.

The basic brown dwarf would be an object with a mass somewhere between Jupiter's mass (318 Earth masses) and seventy Jupiter masses (about 22,000 Earth masses). These objects would have surface temperatures of one or two thousand degrees absolute—about fifteen hundred to three thousand degrees Fahrenheit. Each of them would be slowly cooling, with its contraction yielding progressively less heat with the passage of time. As a result, the relatively young brown dwarfs—those less than a few hundred million years old—would be the "brightest"; that is, they would radiate the most infrared. But none of the brown dwarfs has anything like the luminosity of a star; all of them are at best dim sources of infrared radiation plus a small amount of visible light, that truly deserve their name. How then can we hope to find these objects, assuming that they do exist in large numbers?

## HOW TO FIND BROWN DWARFS

An astronomer who becomes intrigued by the brown dwarf issue has two basic lines of opportunity. He or she may become a theorist who calculates the properties of brown dwarfs—how their sizes, temperatures, and densities change throughout such objects as they age, contract, and eventually become stable, cold bodies. Or he or she may choose to become one of those who hunt for brown dwarfs, eager to help determine how many of these strange objects occupy space in our galaxy. The latter type of astronomer understandably gets more press, the more so as success may be right around the corner.

An astronomer who chooses to search for brown dwarfs has two kinds of brown dwarfs to hunt for. One type consists of the brown dwarfs freely floating in interstellar space, unbound to other stars. This brings us to the basic means to search for *any* type of object: Go out and look for it! Although commendable, such a search plan suffers from the difficulty of finding a brown dwarf all by itself. Brown dwarfs produce relatively little light and not much infrared radiation. Unlike stars, which have nuclear fusion to light their efforts, brown dwarfs radiate energy only because they are contracting. The contraction produces heat, and the heat energy is radiated into space as infrared and a little visible light, but there's not much energy to be radiated.

The second line of attack on brown dwarfs therefore seems better situated to bear quick fruit: Look for brown dwarfs close to real stars. Of course, this does not mean that an astronomer should simply try to *see* brown dwarfs close to their putative stellar companions. Such a search would in fact present greater difficulty, not less, than a search for free-floating brown dwarfs, since the starlight would tend to overwhelm the light from the brown dwarf.

Instead, a search for brown dwarf companions of stars in the Milky Way uses several better thought-out techniques. First, astronomers can look for perturbations that the brown dwarf's gravity produces on the motion of the star that it accompanies. These perturbations may appear in different ways, most noticeably in the changes in position of the star as observed on the sky (which can be carefully measured over several years) and in the velocity of the star along our line of sight. And second, astronomers can look for infrared radiation from the brown dwarf. This amounts to a variant on the basic notion of looking around the sky for what you want to find—a variant that uses the fact that brown dwarfs emit relatively greater amounts of infrared radiation than normal stars do.

## FINDING BROWN DWARFS BY GRAVITY

In Chapter 9 we reviewed the methods that astronomers use to search for planets beyond the solar system. In the search for extrasolar *planets*, it turns out that indirect methods based on the planets' gravitational force seem more fruitful than direct searches for the planets themselves. This is so because planets shine so weakly by reflected light and in infrared radiation that direct searches are hopeless—at least from the Earth's surface.

With brown dwarfs the situation turns out to be the reverse, more or less. Although brown dwarfs have larger masses than planets, and therefore produce larger effects through gravity on their companion stars, they also emit far more radiation (mostly in infrared) than planets do. Brown dwarfs are larger than planets, which means that they emit a greater total amount of radiation, and they are also contracting, which makes them emit still more infrared radiation—the heat released by their slow contraction—than they would if they did not contract at all. In short, brown dwarfs are brown (that is, infrared emitters), but planets are nearly black, hardly emitting anything in comparison to brown dwarfs and stars, shining only weakly in the visible light that they reflect from their parent stars.

Nevertheless, the same techniques that we examined for finding extrasolar planets—looking for the proper motion and radial velocity changes induced by an unseen object's gravitation—will work for brown dwarfs too. Indeed, the star HD114762, where David Latham and his collaborators have found a companion, probably represents the discovery of a brown dwarf, not a planet. A study by Ben Zuckerman, an astronomer at UCLA, of several dozen white dwarfs (the cores of former stars, typically with masses close to the sun's and thus far more massive than any brown dwarf) has revealed several possible companions with masses in the brown dwarf range. Hence astronomers have little trouble pronouncing that brown dwarfs have been located by indirect means in orbit around other stars (or former stars). More exciting still, and a discovery with a recent date upon it, is the fact that astronomers now believe that they have made direct observations of brown dwarfs.

## DIRECT OBSERVATION OF BROWN DWARFS

In the spring of 1989 a team of astronomers led by William Forrest of the University of Rochester announced to their colleagues that they had made, during the previous fall, the first definitive observation of brown dwarfs. The Forrest team decided to search in the region of our galaxy known as the Taurus cloud, or more formally as the Taurus-Auriga star-formation complex. In this region, located where the constellations Taurus the Bull and Auriga the Charioteer intersect, we know that stars are forming by the thousands because we see young stars that have been born there within the past few million years. (Astronomers feel confident that they can assign ages to stars when they see them still partially wrapped within the gaseous cocoons from which they formed.) If the Taurus cloud contains young stars, and if brown dwarfs form along with stars, then the cloud should also contain young brown dwarfs. Since young brown dwarfs are the type of brown dwarf that radiates the most energy and should therefore be the easiest to detect, Forrest and his colleagues felt that they had made the right choice when they spent their energies searching this region.

In hindsight the group appears to have made a good choice. Forrest and his colleagues went to the Mauna Kea Observatory, where NASA's Infrared Telescope Facility provides the best infrared view of the heavens available on the Earth's surface. They studied twenty-five stars in the Taurus cloud and found that eleven of these twenty-five seemed to have companions, which appeared as indistinct dots on the infrared images. The eleven stars with possible companions yielded a total of twenty candidate companions (many stars showing more than one candidate), and of these twenty, nine showed the dark red colors that characterize cool objects such as brown dwarfs (see Photo 23).

These nine stars have yet to be established as definite brown dwarfs. Although some astronomers believe them to be stars whose light has been reddened by its passage through regions rich in interstellar dust grains, the Forrest group is confident that at least some of these nine objects are in fact the long-sought brown dwarfs. To reach this conclusion, the astronomers must use a process of elimination, excluding other possible natures for the objects they have seen. This process forms an essential part of much scientific research; as Sherlock Holmes was fond of saying, "When you have excluded the impossible, whatever remains, however improbable, must be the truth."

The chief way to eliminate possibilities other than brown dwarfdom for Forrest's candidates is to establish that the candidates are in fact not much more distant than the stars in the Taurus cloud. If the objects are significantly farther away than the Taurus cloud, they have luminosities (energy output per second) much greater than those given when we assign them distances equal to those of the Taurus cloud. Then the candidates might be rather dim stars—but stars all the same—seen at relatively large distances.

Thus a distance determination would provide crucial evidence. But measuring the distances to astronomical objects is never trivial; the task provides astronomers with some of their finest moments and sharpest discussions. Distance measurements are absolutely essential if an astronomer hopes to determine one of the most fundamental properties of an object: its *luminosity*, that is, the total amount of energy it emits each second. To determine an object's luminosity you must know two quantities: its *apparent brightness*—how bright the object *appears*—and its *distance* from you. You may be observing a relatively low-luminosity object that lies relatively close by, or you may have in your sights an immensely luminous object at a far greater distance; both situations will yield an object of modest apparent brightness.

In the case of the putative brown dwarfs, a determination of their distance (or distances) would settle the matter completely. *If*—and we must not lose sight of the if—the objects are in the Taurus cloud, and therefore have the same distance from us as the stars in that cloud, then we can calculate their luminosities because we know the objects' apparent brightnesses and distances. These luminosities would be less than the luminosity of any true star, even of the most modest ones, and would prove that we have seen brown dwarfs. On the other hand, if the objects' distances are significantly greater—that is, if they lie far beyond the Taurus cloud, and we simply happen to observe them *through* that star-forming region—then they may well be dim stars and not brown dwarfs at all.

Unfortunately we have no direct way to determine the distances to the candidate brown dwarfs and thus can obtain no definitive answer to the puzzle. But astronomers can do something a bit more subtle that yet may reveal the objects' distances: Measure the objects' proper motions.

As we discussed in Chapter 9, the proper motion of a

star (or brown dwarf) refers to the measured change in a star's position *as we see it on the sky*; that is, it denotes the result of the star's motion with respect to us across our line of sight. It is important to note that in determining any star's proper motion, astronomers are now quite aware of the fact that the Earth revolves around the sun. This revolution produces an apparent back-and-forth motion, called the parallax shift, in the position that we observe for any star during the course of a year. Since astronomers understand the parallax shift quite well, they can allow for its effects. Indeed, this parallax shift provides the basis for measuring the distances to the closest stars—the only stars whose parallax shifts are large enough to be measured with any accuracy.

Through years of patient effort, astronomers have measured the proper motions of the stars in the Taurus cloud. How does this compare with the proper motions of the nine candidate objects? As things turned out, Burton Jones of the Lick Observatory of the University of California had been studying the proper motions of those objects in the Taurus cloud that had been recorded on the photographic plates of the famous Palomar Sky Survey, which mapped the entire sky visible from Palomar Mountain photographically during the 1950s. Of the nine candidate objects, Jones had found six on the Sky Survey plates. Of these six, four showed proper motions consistent with those of the stars in the Taurus cloud. The conclusion followed that these four objects are likely to have the same distance from us as the stars in the cloud—in fact, formed within that cloud—and therefore have the low luminosities calculated to be typical of brown dwarfs.

Four brown dwarfs sounds like a small number, but astronomers are experts in extrapolation. From these four objects, Forrest and his colleagues speculate that the Taurus cloud contains somewhere between ten thousand and a million brown dwarfs. This conclusion rests on the fact that the Forrest group has studied only a tiny fraction of the volume within the Taurus cloud, and therefore a far greater number of objects remain to be found.

And if the Taurus cloud contains something like a million brown dwarfs, how many does the entire Milky Way have? Perhaps a billion, perhaps ten billion, perhaps a hundred billion. It all depends on how representative the Taurus cloud is of the entire galaxy and on how long the brown dwarfs apparently found in the cloud remain visible at the level of apparent brightness sufficient for Forrest to have detected them. But to have passed in a single paragraph from four brown dwarfs in Taurus to a billion in the Milky Way represents a good day's work for any theorist.

Even a billion brown dwarfs—or a hundred billion—would not make a significant contribution to the mass of the Milky Way. Since the Milky Way contains some three hundred billion stars, and since a typical brown dwarf has a few percent of the mass of a typical star, not billions but *trillions* of brown dwarfs would be needed to make a significant contribution to the Milky Way's total mass. This might be the case, but we cannot be at all sure that it is so.

In summary, brown dwarfs have probably been found—four of them by William Forrest and his coworkers, one by David Latham, several (not quite so certain) by Ben Zuckerman. Many more brown dwarfs probably exist in the Milky Way and in other galaxies. But if brown dwarfs make a significant contribution to the mass of the galaxy—and of other galaxies—they do so because a galaxy contains trillions of brown dwarfs. This gnat cannot be so easily swallowed by astronomers, though as we shall see, many of them quite easily swallow the camel of a universe in which the amount of dark matter in unknown form far exceeds the amount of matter that we can see. To some astronomers, brown dwarfs would be a particularly inelegant solution to the riddle of the dark matter. To others, brown dwarfs are entirely acceptable constituents of the galaxy but simply cannot reasonably provide more mass than we find in stars. To a few diehard supporters of brown dwarfs, these failed stars are likely to be *the* basic unit of the universe, with more mass in total than all the stars that shine.

For the general public, the important point about brown dwarfs is that these "failed stars" apparently exist in great numbers. They deserve recognition as a newly discovered component of the Milky Way. Furthermore, brown dwarfs nicely highlight a crucial fact. The process of contraction that forms stars works only if the contracting object has a mass sufficient to squeeze the star to the point where nuclear fusion begins. Take another look at Jupiter and the sun, and rejoice that at least one object in our vicinity had enough mass to set itself on nuclear fire.

# 12
# Exploding Stars and the Seeds of Life

STREWN AMIDST THE Milky Way's stars and star-forming regions, among the planetary nebula shells and the regions lit from within by young hot stars, are a set of expanding gaseous envelopes, wispy remnants of supernovae—stars that have exploded. These supernova remnants have been blown outward toward interstellar space by the force of a titanic eruption, and gradually slow down (from speeds of tens of thousands of kilometers per second) as they encounter interstellar gas and dust. A supernova remnant therefore eventually merges with the overall material floating between the stars in our galaxy. If that were the full story, little more need be said about exploding stars and the matter they eject into space. But there *is* more: These explosions made our planet, our environment, and ourselves. This fact gives supernova explosions a greater claim to our attention and makes it worthwhile to investigate why some stars explode, though most fade away as degenerate white dwarfs.

For centuries, titanic stellar outbursts have borne the astronomical name supernovae. The word *nova* derives from the Latin word for new and is pluralized with an e. The name signifies that a supernova typically appears as a new star where no star was seen before. Mere novae likewise appear as new stars, on closer inspection they turn out to be rather ordinary stars that flare up repeatedly—though not to the point of self-destruction—as they age. But supernovae are far more intrinsically luminous than novae, and all of the supernovae ever observed by mankind, with a few significant exceptions, have come from stars that were too dim to have been seen before their explosions.

## THE TWO TYPES OF SUPERNOVAE

A supernova blows just once, and represents the last dying gasp of a star. Oddly enough, such a final curtain call can occur—so astronomers now believe—in two different ways. Furthermore, the two different types of supernovae are roughly equal in numbers: Astronomers detect about as many Type I as Type II supernovae.

A Type I supernova typically occurs when one member of an evolving double-star pair has become a white dwarf—the core of a star whose outer layers have evaporated into space and which has ceased nuclear fusion entirely. Consider what happens when the stellar companion to the white dwarf becomes a red giant (as all stars will). A red giant is an aging star that puffs its outer layers to enormous size, surrounding its contracting core with a huge rarefied envelope of hydrogen and helium gas. If the red-giant companion swells to the point that part of its outer layers rain onto the white dwarf's surface, the white dwarf will steadily accumulate matter from its companion, building a new surface on top of its crystalline structure of carbon nuclei and electrons (see Figure 25).

This newly arriving matter comes from the outer layers of the red giant, which means that unlike the matter in the red giant's core, the infalling material contains plenty of hydrogen atoms. The red giant's core has processed all its hydrogen into helium and will soon begin (if it is not doing so already) processing the helium nuclei into carbon. The star's outer layers are another story. Since nuclear fusion occurs only in the core, and since mixing of the outer layers with the inner layers occurs only partially—if at all—the outer layers remain much as they were when the

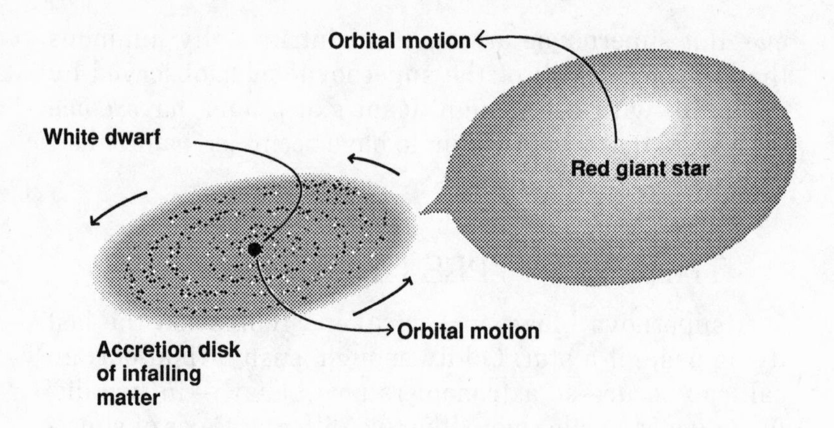


Figure 25: A Type I supernova arises when matter from a bloated red-giant star falls onto a companion white dwarf. The material accumulates at a high temperature, then suddenly fuses to produce an explosion. *Drawing by Crystal Stevenson*

star formed. The layers are therefore hydrogen-rich, like the rest of the universe until something happens to it.

The hydrogen-rich loam settles onto the white dwarf, bit by bit, year by year, steadily accumulating a thicker layer around the hot surface. As the matter accumulates, the pressure induced by the white dwarf's gravity squeezes the material to higher densities and higher temperatures. The *surface* of the white-dwarf-with-hydrogen-wrapping now mimics the *interior* of an ordinary star: Material rich in protons (hydrogen nuclei) is compressed to ever-increasing density and temperature. The same fundamental result occurs in the two cases: At temperatures of about ten million degrees Celsius the protons begin to fuse together.

In an ordinary star this fusion soon reaches a steady rate and continues for billions of years because the star has a natural self-balancing act. The star's gravity tends to compress the star, while the energy released in nuclear fusion tends to explode it. The star owes the balance that arises between these two tendencies to the fact that nuclear fusion and energy release occur at its center. The star's outer layers wrap and compress that center in a protective blanket thousands of kilometers thick. If the star creates too much energy through nuclear fusion, the additional energy expands both the core and the compressive blanket slightly and thereby reduces its rate of nuclear fusion to return to its normal size.

In contrast to this self-regulation through feedback, a white dwarf with hydrogen-rich matter on its surface stays calm until it blows completely apart. When a white dwarf accumulates a blanket of material from its bloated companion, and when protons in that blanket start to fuse, nothing exists either to inhibit the fusion or to readjust the star, because nothing significant lies outside the region where nuclear fusion occurs. As the star's entire envelope reaches a temperature close to ten million degrees, the entire layer of material suddenly fuses.

This fusion amounts to a hydrogen bomb about $10^{30}$ (a million trillion trillion) times greater than the mightiest fusion weapons exploded on Earth. When that bomb goes off, it blows the star—outer envelope, white-dwarf interior, and anything else the star may have up its sleeve—to kingdom come. According to astronomers' calculations, a white-dwarf supernova—called a Type I—leaves nothing behind of the star and must have a frightful effect on its companion, although some of the companion may be left after the explosion.

Type I supernovae therefore have a certain air of finality: You begin with a white dwarf and end with memories of one mighty explosion. In that sense, Type II supernovae may be called the modest type, since (again according to calculations) they typically leave something behind other than the material blasted outward by the explosion, or the companion star only partially annihilated by the white dwarf's outburst. Still, Type II supernovae need not—and do not—hide their light under a bushel. They too make quite a glow.

## TYPE II SUPERNOVAE

Type II supernovae do not require a companion star; they occur in single stars whose cores collapse. To understand why this collapse occurs, we must recognize the key fact—completely unrelated to intuition—about white dwarfs: They can exist only up to a certain maximum amount of mass, but not beyond.

White dwarfs hold themselves up against their own gravity through the exclusion principle, which describes the refusal of electrons to pack together beyond a certain density. Once the electrons refuse any tighter packing, they in turn support the carbon nuclei against further contraction through the mutual electromagnetic attraction of negative charges (electrons) for positive charges (carbon nuclei). All this occurs when the density of matter has risen to about a million times the density of water, and the degenerate matter (matter in which the exclusion principle plays an important role in the bulk behavior) simply resists further compression with a steadfastness that would do credit to a martyr. But the same rules of quantum mechanics that describe how this works also make a straightforward statement, first elucidated by the Indian-American astrophysicist Subrahmanyan Chandrasekhar: No degenerate matter can exist in a stable state if its mass exceeds 1.4 times the mass of the sun.

Although one can intuitively sense that the exclusion principle cannot do everything (that is, it cannot hold up an infinitely large amount of mass against its own gravity), no intuitive explanation exists for why the limit should be 1.4 times the sun's mass. Furthermore, our intuition can provide only a modest guide as to what happens if and when a star with more than this mass limit runs out of ways to produce the energy to fight its own gravity. Intuition tells us the star will collapse, but the details belong to the astronomers who have spent more than thirty years at work on the problem of stellar aging and stellar death. Their calculations show that in the minority of stars—those with more than 1.4 times the sun's mass by the time they have ended their red-giant stages—the core will indeed collapse, and that this collapse will occur when the star's center has become mostly iron.

Iron nuclei typically each consist of twenty-six protons and thirty neutrons. They are thus much larger than protons of helium-4 nuclei (two protons and two neutrons), but much smaller than, say, lead-206 (82 protons and 124 neutrons) or uranium-238 (92 protons and 146 neutrons per nucleus). What distinguishes iron from still heavier nuclei is this: Nuclear fusion that makes iron, or makes nuclei lighter than iron, *releases* additional energy of motion, whereas nuclear fusion of iron nuclei to make still heavier nuclei *consumes* energy of motion. In other words, to fuse nuclei up to and including iron releases kinetic energy, but to fuse iron consumes kinetic energy.

From a star's point of view, this means that iron marks the end of the long road of producing energy of motion to oppose self-gravitation. Until a star's core becomes mostly made of iron nuclei, the star has a fighting chance—which it seizes—to keep from collapsing, because the fusion reactions keep on making new energy of motion to share among the star's particles. The furious motion of these particles creates the pressure that the star needs to avoid gravitationally induced collapse. But once the core becomes mostly iron, the party is over, for no source of new kinetic energy exists (see Figure 26). Gravity wins, so collapse occurs. The collapse of the core requires about a second, and has serious repercussions.

In the second of collapse, the different parts of the star's core rush at one another in amazing fury. (The star's outer parts rush toward each other too, but since they have farther to go, they never reach each other before an explosion blasts them into space.) In this furious inrush, nuclei collide so violently that some fusion *does* occur—the result of gravity pulling the star violently together—but most of the collisions strip the nuclei apart into protons and neutrons. Thus the collapse allows the star to undo, in nuclear-fusion terms, what took millions upon millions of years for the star to accomplish.

Once the protons and neutrons have been "dissolved" from larger nuclei, the continuing inrush makes protons

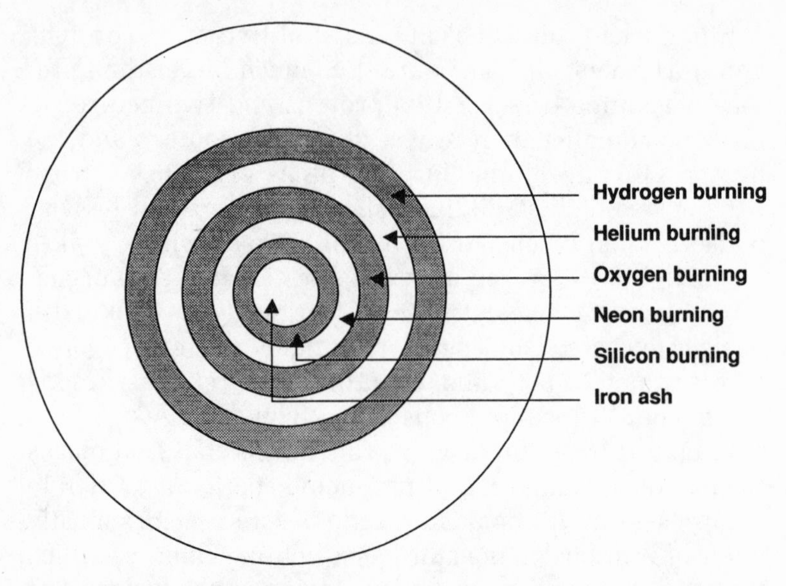


Figure 26: As a star ages, it fuses progressively heavier nuclei closer to its center. Since iron nuclei do not release additional kinetic energy upon fusion, iron marks the end of this process, and the star's core must collapse once it has become mostly iron. *Drawing by Crystal Stevenson*

fuse with electrons (which are present in every star in large quantities) to produce neutrons and neutrinos. Protons and electrons will fuse only under the influence of weak forces, which as their name implies are far weaker than the strong forces that govern ordinary nuclear fusion, such as the hydrogen-to-helium reactions that make stars shine. Hence protons and electrons fuse only in the most extreme circumstances—for example, those within the collapse of a star's core.

## COMPRESSION FORMS A NEUTRON STAR

The combined effects of the stripping out of protons and neutrons plus the fusion of protons with electrons yields a mixture of neutrons and neutrinos. The neutrinos depart the scene at the speed of light, helping to blow off the star's outer layers as they do so. The neutrons stay behind, and in most cases (relying, once again, on astronomers' calculations) will form a neutron star, an object made entirely of neutrons—not really a star, but rather a former stellar core whose collapse has turned it into a neutron crystal.

The neutron star keeps from collapsing further through our friend the exclusion principle. Just as the exclusion principle prevents the electrons in a white dwarf's carbon-electron mixture from being squeezed any more tightly, so too in the simpler situation of a neutron star (made only of neutrons), the exclusion principle acts on the neutrons and in effect tells them that they will squeeze just so far and no more. The squeezing does compress the material to a simply stupendous density, about $10^{14}$ (one hundred trillion) times the density of water, so that a teaspoonful of neutron-star material brought to Earth would weigh as much as the *Queen Mary*. But the squeezing has to stop, and the neutron star can maintain itself indefinitely with a diameter of about twenty miles. Twenty miles! An entire stellar core squeezed to the size of Manhattan! Such is the fate of stars with large masses—the ones that burn themselves out far more rapidly than the sun.

The core's fate may be to become a neutron star, but the star's outer layers have other work in store. As the outer layers fall inward, deprived of their usual support by the core's collapse, they eventually meet the surface of the newly formed neutron star. The kinetic energy of their inrush compresses the neutron star to a somewhat smaller size than it cares to have, and then, like an overtight rubber ball, the neutron star "bounces" to a slightly larger size. This bounce suffices to blast the star's outermost layers into space at high velocity. The bounce starts a shock wave—a sudden transition to a higher density and temperature—propagating outward from the neutron star through the surrounding material. The material close to the neutron star moves outward only sluggishly (by astronomical standards), but because the density of material grows less and less as the shock wave passes outward, the shock accelerates the material with greater and greater ease. As a result, the outermost few percent of the star blast outward hundreds of times more rapidly than our fastest rockets can travel, and a tiny fraction of the originally inrushing material, perhaps one part in a billion, accelerates to speeds greater than 99 percent the speed of light.

And there you have a Type II supernova: The star's core has collapsed to form a neutron star, and this collapse triggers an outwardly accelerating shock wave that explodes the star's outer parts. We can't see the core collapse (save by the neutrinos described below), but we can and do see the outward explosion, whose light and other forms of electromagnetic radiation spread outward from the scene at three hundred thousand kilometers per second. In a large galaxy such as our Milky Way, a Type II supernova occurs about once in every century. Many of these supernovae remain hidden from our discovery because the interstellar dust that lies in our galaxy's midplane blocks their light completely. This interstellar absorption represents quite a feat by interstellar dust, since an unblocked supernova shines for a few weeks with about a billion times the sun's luminosity.

## THE CRAB NEBULA AND THE SUPERNOVA OF A.D. 1054

The most astronomically famous remnant of a supernova explosion appears in the constellation Taurus and has been conclusively identified as the remains of a star whose supernova demise was observed on Earth on July 4, 1054. (Of course, the star had actually exploded long before—about five and a half millennia before, according to our best estimates. All of astronomy is out-of-date news, history rather than a view of the present.) Those observations were recorded by Chinese historians, by at least one Islamic scholar, and apparently by the Anasazi of what is now the American southwest. What all the observers saw was a new star where none had been seen before, a star

Photo 19: A globular cluster consists of half a million or so stars held together by their mutual gravitational attraction, and forms one of the original subunits of the Milky Way. *Lick Observatory*

Photo 20: If a planet the size of Jupiter moved around Alpha Centauri in an orbit the same size as Jupiter's orbit around the sun, our best telescopes could not see the planet in the glare of light from the star. The "spikes" arise in the telescope. Alpha Centauri is in fact a double star. *Photograph courtesy of National Optical Astronomy Observatories*

Photo 21: The Hubble Space Telescope, launched on April 24, 1990, contains the finest guidance system ever sent into space—and a mirror that needs correcting. *NASA*

Photo 22: The first serious attempts to detect gravity waves were made during the late 1960s at the University of Maryland by Joseph Weber, who hung thick aluminum rods and monitored them with extreme care. *Photograph courtesy of Joseph Weber*

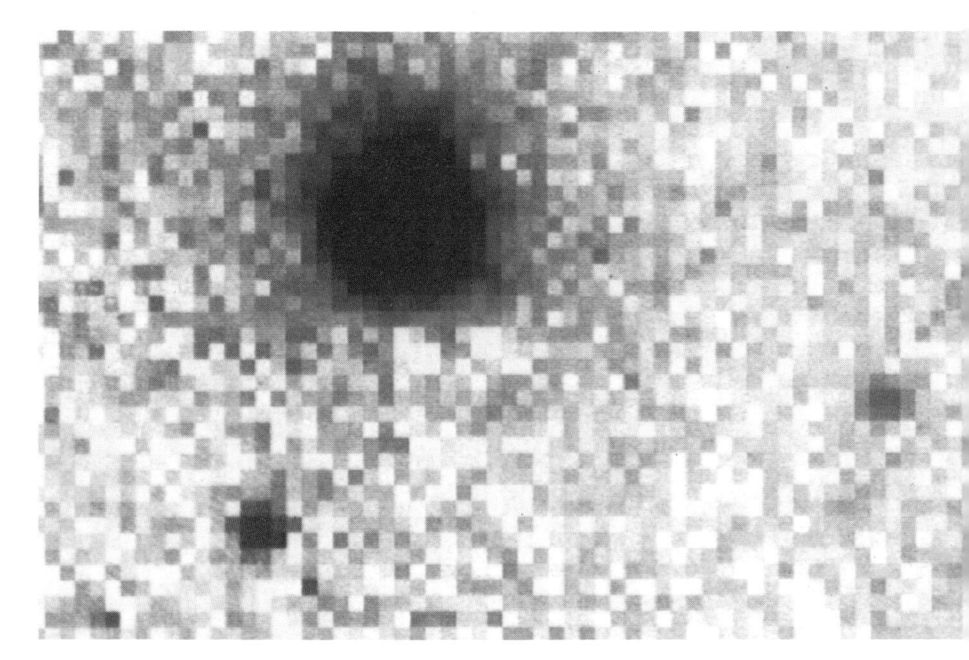

Photo 23: One of the brown-dwarf candidates found by William Forrest and his colleagues appears as a rather nondescript object at the right center of this infrared photograph. *Photograph courtesy of the University of Rochester*

Photo 24: The Crab Nebula, the remnant of the supernova of A.D. 1054, is an expanding web of gas some 5,500 light-years away in the direction of the constellation Taurus. *Lick Observatory*

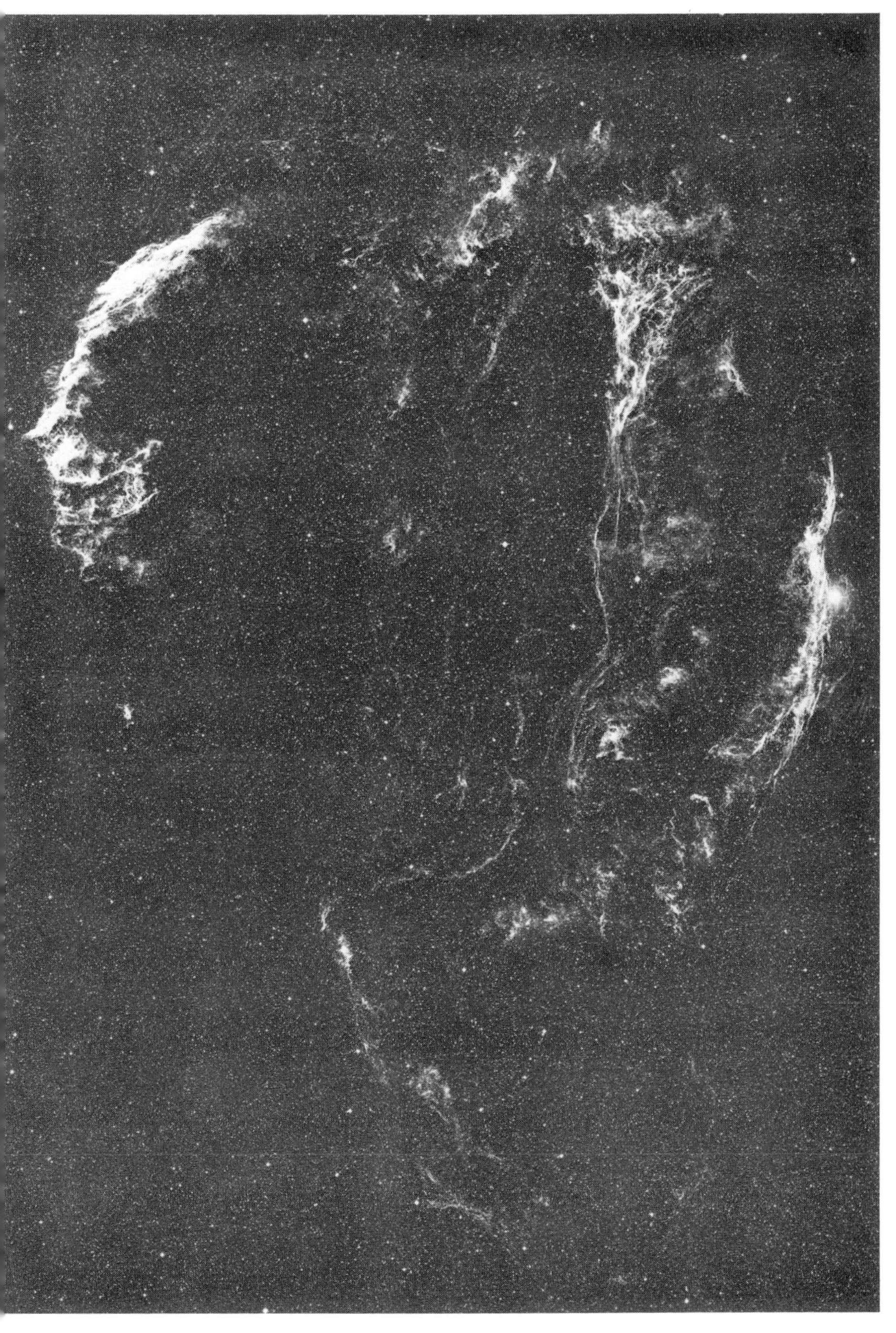

Photo 25: The Veil Nebula, a supernova remnant about twenty thousand years old, gives some indication of the way that supernova-made elements gradually merge with the rest of the interstellar matter in the Milky Way. *Hale Observatories*

Photo 26: Supernova 1987A (bottom) appeared close to the Tarantula Nebula in the Large Magellanic Cloud, an irregular galaxy that is a satellite of the Milky Way and about one hundred sixty thousand light-years from the solar system. *Photograph courtesy of the Royal Observatory, Edinburgh*

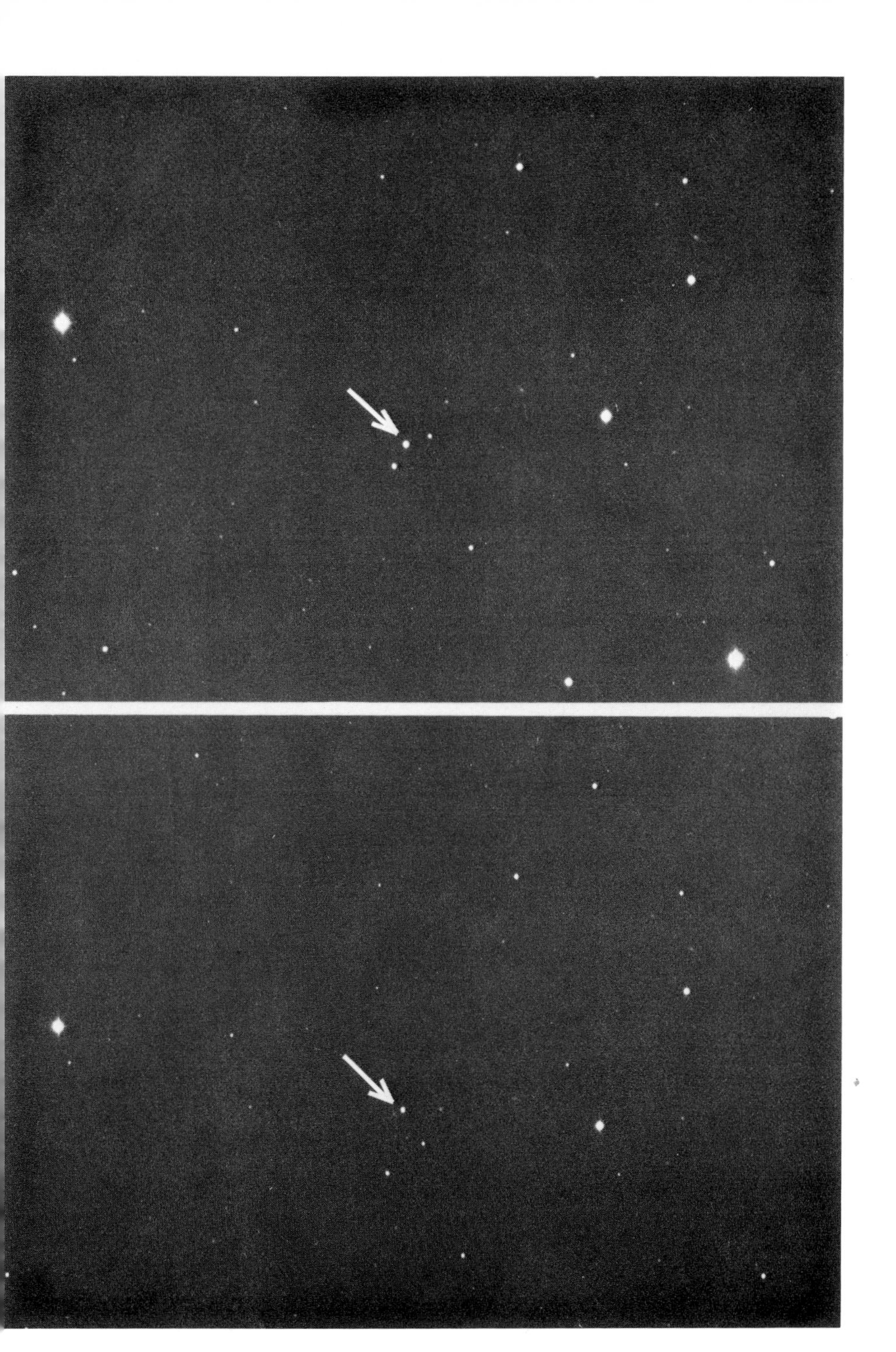

Photo 27: Clyde Tombaugh discovered Pluto in 1930 by noting tiny changes in its position on the sky caused by Pluto's orbital motion during several months. Note how dim Pluto appears in comparison with even faint stars. *Lick Observatory*

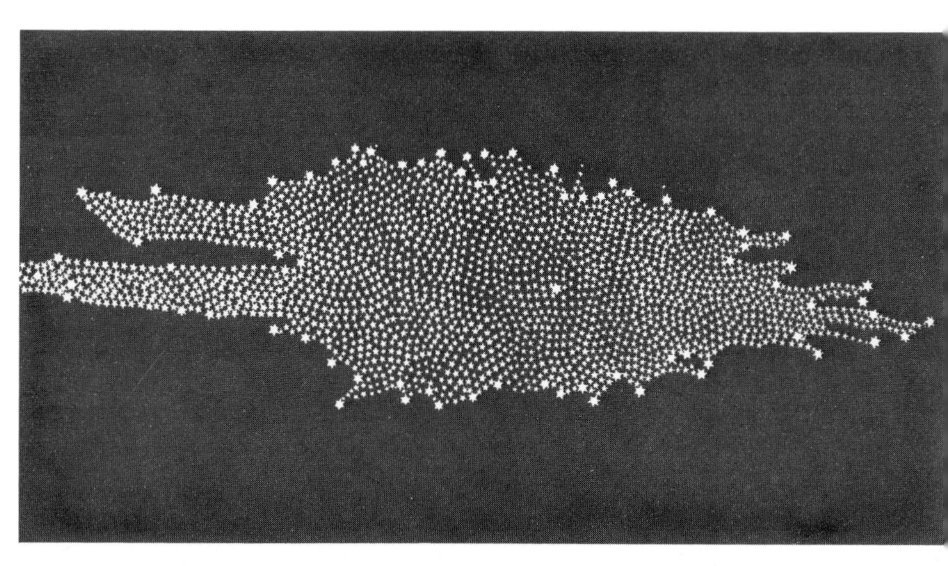

Photo 28: Sir William Herschel's drawing of the Milky Way, made during the 1780s, was based on his star counts in various directions. The drawing shows a somewhat flattened form not much like the actual disklike shape of our galaxy. *Courtesy of Yerkes Observatory*

that within a day became the brightest star in the sky and held this title for several weeks, gradually fading in luminosity and disappearing from view after about six months.

This was a supernova, and a good one, but we would pay it relatively little attention were it not for the supernova remnant that we now observe. The Crab Nebula owes its name to its supposed resemblance to a crab (see Photo 24). Totally invisible without a telescope, the nebula was first described and recorded by the English astronomer John Bevis early in the seventeenth century. Charles Messier made the Crab Nebula number one in his famous list, and it has had a special place in astronomers' hearts in recent years, for the Crab Nebula is nearly the closest and by far the best-studied remnant of supernova explosions.

Seen in visible light, the Crab Nebula shows a complex filamentary structure within which more diffuse emission appears. More careful study has revealed an important distinction between the filaments and the diffuse light. The former radiate only at certain visible-light wavelengths, most notably at the wavelengths typical of hydrogen atoms that have been excited by collisions or photon impact to a higher energy state. When the atom drops back into a lower energy state, light emerges with characteristic, sharply defined wavelengths. Hence the filaments seem to consist primarily of hydrogen gas that has been excited by a large input of energy believed to result from the explosion of this gas away from the supernova.

The diffuse light arises from an entirely different situation, the synchrotron process described previously. The synchrotron process produces light only when particles have been accelerated nearly to the speed of light and testifies to the violence involved in the explosion. Hence the Crab Nebula—the first cosmic object in which the synchrotron process was positively identified—bears the mark of cosmic violence.

Dozens of supernova remnants have been identified in the Milky Way, but they are all older and milder than the Crab Nebula, with the possible exception of the radio and x-ray source Cassiopeia A (see Photo F). The Veil Nebula (see Photo 25) resembles what the Crab Nebula may look like in another twenty thousand years and gives a good impression of what happens to the material blasted from supernovae: It gradually merges with the rest of the interstellar gas and dust, ready someday to form a new generation of stars.

No one alive has seen a supernova in the Milky Way: The last sure supernova appeared in the year 1604 and the last suspected supernova in the year 1687. (The possible supernova of 1687, whose remnant is now Cassiopeia A, lies in a region of high but probably not complete absorption by interstellar dust.) Statistically, even allowing for interstellar absorption, we're due. And as it happens, we *do* have a recent nearby supernova to examine, one not in our galaxy but in the Milky Way's closest neighbor galaxy, its satellite called the Large Magellanic Cloud. We're talking about the big supernova news of this century, Supernova 1987A.

## THE GREAT SUPERNOVA OF OUR TIME

On the night of February 23, 1987, a Canadian astronomer named Ian Shelton, working at the Las Campanas Observatory in Chile, noticed a new bright star that had appeared close to the main part of the galaxy called the Large Magellanic Cloud. (Not everyone could have done this: The skies at Las Campanas, chosen for the clarity of the dark night, are filled with stars, enough to seem overwhelming to those who no longer trouble to learn which stars are which. Shelton's new star was "only" one among the hundred or so brightest in the sky.) Shelton consulted with his fellow astronomers and realized that in all likelihood a supernova had appeared. The news went by car to La Serena, the Chilean town close to the observatory, and by radio and telephone around the world. Soon all astronomers—not to mention the general public—knew that Supernova 1987A had burst forth as the first supernova seen in the Local Group of galaxies in a hundred years and the first to appear in the Milky Way or its satellite galaxies since the seventeenth century (see Photo 26).

Several hours before Shelton's discovery, a burst of neutrinos (about two dozen in all) had been recorded in neutrino detectors located deep underground in Japan, the Soviet Union, and Ohio (see Photo G). Had these neutrino detectors been equipped with more sophisticated real-time analyzers as well as an alarm system to alert the physicists who govern them, the physicists could have stolen a march on the astronomers and announced a possible supernova seen by its neutrinos. Since neutrinos are massless particles that travel at the speed of light, and since the neutrinos from the supernova outburst had several hours' head start on the light (which was temporarily trapped within the supernova's expanding outer layers), the physicists had the time advantage, but one that they did not seize—not this time. After all, Supernova 1987A was the first object beyond the sun to be detected by its neutrinos! Neutrinos are so reluctant to interact with matter that nearly all of them pass right through the Earth with no effect whatsoever. In fact, each person on Earth received a bombardment by a million or so neutrinos from the supernova. Astrophysicists were delighted to find the pulse of neutrinos from an exploding star. They praised the serendipitous fact that the "neutrino detectors" had been built not to find neutrinos but to search for an only distantly related phenomenon, the possible decay of protons (particles that form part of every atomic nucleus) into other particles. As it turned out, a proton-decay detector also makes a fine neutrino detector, but in all honesty the scientists running the detectors weren't thinking much about neutrinos—until their experiments became famous for detecting them.

The neutrinos played a key role in reassuring astronomers that their models of supernova explosions, derived from exhaustive computer studies that attempted to incorporate every significant parameter known about how nuclei fuse together, made sense. These models predict that the collapse of a stellar core must, as we have seen, release copious streams of neutrinos from the reactions among nuclei that occur during the collapse. For a brief second or so, the core is so dense that not even neutrinos can easily escape from it. But then, after a second or so, the combined effect of the energy released in the "bounce" from the neutron star and the neutrinos' tendency to rush outward blasts the supernova's outer layers into space. Once the layers thin out a bit—to a density merely a few thousand times denser than water—the neutrinos can pass through them like a bullet through air. They escape from the scene of their creation, carrying to the universe the news that another stellar core has collapsed. As the explosion continues, the star's outer parts continue to become more rarefied until the light and ultraviolet radiation produced in the heat of the explosion can also escape, adding its (somewhat delayed) news to that of its precursor, the neutrinos.

The observations fit well with the theory: About as many neutrinos were captured by the detectors as were expected from a collapsing star one hundred and sixty thousand light-years away—the distance of the Large Magellanic Cloud. Astronomers barely paused to recall that they were observing a star death that occurred at a time when our ancestors had yet to discover the joys of agriculture and city living. To them the excitement was *now*, and they rushed to observe the supernova in detail. Their activities included ultraviolet observations from the *International Ultraviolet Explorer* (*IUE*) satellite, optical observations from a host of observatories south of or close to the equator (for the Large Magellanic Cloud is invisible from the continental United States or Europe), balloon-borne gamma-ray observations, infrared observations from the Mauna Kea Observatory in Hawaii, and x-ray observations from Soviet and Japanese satellites. The United States, sad to relate, had no x-ray satellite in orbit, though its nearly worn-out solar-observing satellite, the

*Solar Maximum Mission*, did make the first detection from the supernova of gamma rays that penetrated the sides of the gamma-ray detector.

The observations generally confirmed the theoretical picture: The core of a massive star had collapsed, producing a Type II supernova. Just prior to the collapse, the star's central regions had become mostly iron, and the exploded material contained great amounts of iron, cobalt, and nickel nuclei. (Cobalt and nickel are the nuclei just heavier than iron.) The key source of heat within the supernova came from the radioactive decay of a particular type of cobalt nucleus (cobalt-56) into nuclei of iron-56. Each nuclear decay produced a gamma ray, a high-energy photon that collided with surrounding particles, passing its energy on to them and thus heating the entire expanding shell of gas. From the luminosity of the supernova, astronomers could calculate that a mass equal to 7 percent of the sun's mass had been converted into nuclei of cobalt-56, since it would take this amount of the radioactively decaying nuclei to produce the supernova's glow.

Eventually, as the ejected material thinned out still more, x-rays from the hot gas could escape directly into space rather than being blocked by the outward-lying material. Astronomers detected these x-rays about six months after the explosion, in the fall of 1987. By this time, as the first conferences about the supernova came to a happy conclusion in Europe and Virginia, the theorists and observers felt that they had satisfactorily solved the most unexpected aspect of Supernova 1987A: The fact that in this case it was a blue star that blew.

Supernova 1987A was the first supernova with a well-studied precursor star. Generations of astronomers had photographed the Large Magellanic Cloud for reasons quite unrelated to whether any stars there were ripe for explosion. Nicholas Sanduleak had made a list of blue stars that interested him, and as astronomers examined the supernova's location and searched their old plates they saw that the exploding star had previously been known as

Sk −69° 202. This affectionate name honors Sanduleak (Sk), the star's declination (an astronomical coordinate, −69 degrees in this case, that specifies the star's location with respect to the celestial poles), and 202 (the star's place on the list among those with a declination of −69 degrees). Old 202, as the star was often called, had blown its topmost parts. This was surprising in light of the fact that the theory called for a star to explode at the end of its *red*-giant phase, when the star was expanding its outer layers even as its core contracted toward ultimate collapse.

The theory required modification. Before long theorists had demonstrated that although most Type II supernovae arise within red-giant stars, in some cases a red giant can shed its outer layers, leaving behind a smaller, bluer star. The bluer color arises from the fact that the temperature within a star rises steadily from the surface toward the center. Hence the loss of the outer layers reveals the hotter layers below. Only at this point had old 202 blown. In fact, the theorists pointed out, this explained why Supernova 1987A failed to grow as bright as first predicted. The supernova never ranked among the brightest stars in the night sky, as would be expected for an ordinary supernova in the Large Magellanic Cloud. (Think of what this implies for supernovae. The distance of one hundred sixty thousand light-years to the supernova exceeds the distance to nearby stars by ten thousand times. Since any object's apparent brightness decreases as the *square* of its distance, an "ordinary" supernova must have an intrinsic luminosity a hundred million [ten thousand squared] times greater than that of the bright stars close to the sun.) Instead for a few weeks Supernova 1987A merely ranked as one of the few dozen brightest stars in the sky and has then slowly faded away as the years have passed.

Supernova 1987A's failure to grow so bright stemmed from its compact size, which in turn related to its blue color, when it blew. Since the exploding star was smaller than usual, the explosion expended more energy than usual in blowing off the star's outer layers and had less energy left over with which to shine. Once again the theorists had an explanation; they are now ready for the next blue star that becomes a supernova.

Time passed, and as the supernova theorists and observers debated points of great interest concerning this celestial marvel, a certain amount of energy went into waiting for the pulsar. As we discussed in Chapter 8, when a stellar collapse produces a neutron star, astronomers expect that in many cases the neutron star will produce a pulsar, a source of radio (often of optical) emission that emerges in regularly spaced pulses. The regular spacing derives from the rotation of the neutron star, and this rotation should begin at an extremely rapid rate (hundreds of times per second) and slow down gradually.

Astronomers took their equipment—machines capable of analyzing light thousands of times per second to look for any repeating pulses—to the Chilean observatories. They made radio observations of the supernova remnant with analogous equipment, obtaining sufficient data on tape to make the mind reel. And then, early in 1989, a group of scientists led by Jerome Kristian of Caltech and Carl Pennypacker of the University of California, Berkeley, announced that they had found regular pulsations in the light from Supernova 1987A on just one night. These pulsations had recurred nearly *two thousand times* per second. This frequency, though startlingly high, seemed barely possible to theorists who predict how neutron stars produce pulsars. Luckily for the theorists but a bit unsettlingly for the observers, it turned out some two years later that the Chilean observations had detected not the pulsar from the supernova but a high-frequency emission from the electronic system used to search for pulsars. In other words, the astronomers had detected their own equipment—neither the first nor the last time such an error has occurred in this high-tech world.

The fact that we still have found no pulsar from Supernova 1987A has not caused astronomers to stop looking for one. Because a supernova ejects clouds of material, and because this matter can flow between ourselves and any pulsar left behind, it is quite possible that—at least during the first few years after the explosion—we will obtain only fleeting, if any, chances to look directly into the explosion's center and thus to observe the pulsar. The pulsar may be pulsing its heart out, even though matter between there and here may be blocking all hope of seeing it at work in its infancy. Furthermore, the pulsar may indeed be emitting pulses of light and radio waves but not in our direction. That is, the pulsar's "lighthouse beam" may sweep past other observers but never over the solar system. Pulsar hunting can prove disappointing at times, though even from negative results astronomers can learn such intriguing facts as the number of stars that explode without producing a pulsar that we can detect.

## SUPERNOVAE AND THE SEEDS OF LIFE

Meanwhile, we may note an absolutely fundamental aspect of supernovae: their effect on the evolution of the universe. Supernovae are the source of most of the elements other than hydrogen and helium that we find around us, most notably in our planet Earth and in ourselves. Some of these elements, such as carbon and nitrogen, come in part from stars that gently expel the elements they have made into space during their red-giant phases, toward the end of their lifetimes but well before some of these stars become supernovae. But for the bulk of the heavy elements—oxygen, magnesium, silicon, nickel, iron, lead, gold, and uranium, for example—if you ask, What made them?, the answer is: Stars that exploded.

Stars that exploded as supernovae have made new elements in two ways: gradually and all at once. The elements up to and including iron and nickel were made *before* the star exploded as part of the late stages of its presupernova evolution. In contrast, stars do not make any of the elements heavier than iron and nickel *until* they explode. Then, in the brief inrush and ejection of material, temperatures temporarily rise so high that nuclear fusion can occur, and a small amount of elements such as silver, mercury, and gold appears. This difference between slowly and quickly made elements helps explain why the latter are so much rarer than the former. When you pan for gold, you are seeking atomic nuclei fused in the first second of a supernova outburst.

All of the supernova-made elements—the oxygen and silicon, the iron and nickel, the palladium, titanium, vanadium, chromium, and zinc—percolate through the Milky Way, mulching our galaxy with a rich cosmic loam. Over time, supernova explosions (with some contribution from red-giant outflows) have enriched the original mixture, which essentially consisted of only hydrogen and helium. Today, ten billion years after the galaxy formed, about 1 percent of the mass in stars consists of *all* the elements other than hydrogen and helium. This may not seem like much, but it makes a difference—not to stars, which run perfectly well without any of these elements, but to ourselves.

We live on a planet that formed close to its star some four and a half billion years ago. Once the sun began to shine, its heat evaporated all the hydrogen and helium, the two lightest elements, in its immediate vicinity, except for the great bulk trapped within the sun by its gravity. Hence our Earth and the three other inner planets represent the residue left behind once the majority of the original material had escaped into interstellar space. We live on the husk, so to speak, of the cosmic ear that once occupied our space. This husk contains relatively little hydrogen (the oceans, which have plenty of it, form only the tiniest fraction of our planet's total mass). Most of the Earth consists of oxygen, silicon, iron, and aluminum, four elements basically made in supernovae—stars that exploded long before the sun and its planets began to form.

Like our planet, living creatures on Earth also consist largely of star-made elements. The four most abundant elements in life-forms on Earth are carbon, hydrogen, nitrogen, and oxygen, and we could not live without additional elements such as phosphorus, iron, and zinc. Thus we as individuals largely consist of star-stuff, matter processed inside stars that later exploded. We should regard supernovae not merely as cosmic extravaganzas but as a key link in the cosmos that left us here to admire it.

# 13

# Most of the Galaxy Is Missing

Stars are bright, and the Milky Way is a galaxy of brightly shining stars. This once seemed self-evident, but to the chagrin of astronomers, recent studies have revealed that most of the Milky Way, perhaps 90 percent of its mass or more, produces too little light for us to see it. Billions of stars make up only a small fraction of our galaxy, composed of we know not what, distributed in a way we don't fully understand. The problem, dubbed the "dark matter problem," is not unique to the Milky Way. But to live within a hidden galaxy, to have its basic parts so close and yet invisible, has provided a special puzzle to astronomers.

## RETURN TO PLANET NINE

At the turn of the century eight planets were known, along with scores of comets, many minor planets, and myriad moons. But something was wrong. The outer planets, particularly Neptune and Uranus, didn't follow the orbits they should have. Something—perhaps another planet—was perturbing them, nudging them away from the celestial paths gravity dictated.

Astronomers could easily see the problem by taking several photographs, spaced a decade apart, of the positions of the outer planets. Comparing the observed positions with the positions stable orbits would predict, it was clear that something was wrong: The positions did not match. A mystery planet, Planet X, might exist that, although invisible, could affect the paths of the other planets. Astronomers could even calculate the approximate mass and location that Planet X must have to explain the observed positions of Uranus and Neptune.

Planet X is today called Pluto. When it was finally discovered by Clyde Tombaugh in 1930, its position was surprisingly close to that which the calculations—sight unseen—had provided (see Photo 27). However, Pluto's mass, and thus its gravitational effect on the orbits of the outer planets, turned out to be much smaller than the calculations had suggested. In a sense, Pluto was found by accident, since it is not the planet (in terms of mass) that had been predicted. Nevertheless, Pluto provides a good example of how gravity's pull can be used to look for mass that doesn't emit any significant amount of light.

Pluto is the least massive planet, with less than a tenth the mass of the moon. It certainly is not a bright object; at its vast distance from the sun, the feeble light it receives offers only a perpetual, dim twilight. Pluto's small size and great distance make its reflected light remarkably weak, so the planet is visible only with moderate-sized telescopes. But Pluto can influence other objects with even its modest gravity.

Finding planets by the effects of their gravity offers an example of how to search for dark matter. But the bit of dark matter we call Pluto is just a bit of fluff, a speck of matter in comparison with the giant Jupiter or the colossal sun. Pluto as dark matter still makes up only a tiny fraction of a thousandth of 1 percent of the mass of the solar system. But the fact that we could detect, even without seeing, such a minute object testifies to the power of motion as a tracer of mass whether or not that mass produces light.

## TRACING OUT THE MILKY WAY

Not every motion in the galaxy is caused by gravity, but gravity makes everything move. As with the case of Pluto, the motion induced by gravity can be used as a way of tracing out the structure of the Milky Way.

An early and most imprecise way to map the structure of our galaxy counted stars in various directions on the sky. Sir William Herschel of England did this during the late 1700s to derive a crude picture of the Milky Way in which the sun lay near the center (see Photo 28). But Herschel did not know that vast clouds of gas and obscuring cosmic dust hide huge numbers of stars. And this method of counting stars, though noble and simple in execution, could not reveal distances, and hence could not show the breadth of the galaxy.

As discussed in Chapter 9, one of the methods to determine distances relies on the parallax effect—a change in the perspective of stars' positions arising from the Earth's changing position during its orbital year. However, since only a few hundred stars are close enough to exploit this technique, only a tiny fraction of the galaxy's size and structure can be probed in this way. But another way to estimate distances relies on finding stars of known inherent brightness and then uses those brightness characteristics to derive distances.

One important class of stars used in this method are the Cepheid variable stars. The cyclical changes in the apparent brightnesses of Cepheids indicate their inherent brightnesses or luminosities (see Figure 27). Once their periods reveal the actual luminosities, we can compare these luminosities with the Cepheids' observed brightnesses. This comparison tells us the distances to the stars, since we know that any star's apparent brightness decreases as one over the square of the distance to the observer.

In addition to the Cepheids, other types of stars and gas clouds provide "standard candles," classes of objects whose luminosities or sizes astronomers have determined as representative of the entire class. They observe these known quantities scaled by the distance. Using these methods, astronomers have been able to find the size and shape of the Milky Way. Our galaxy is a pancake, a flat

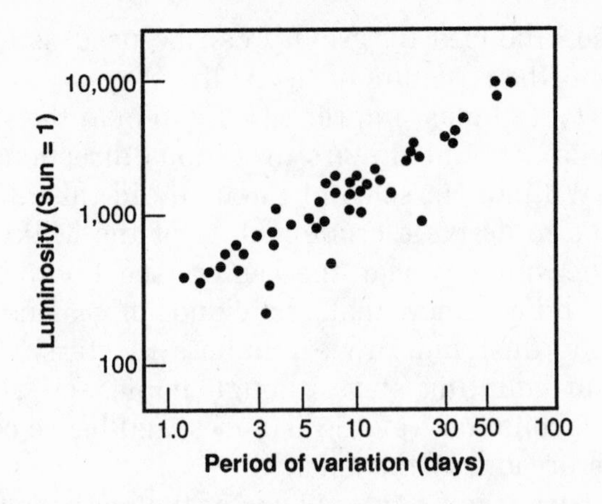


Figure 27: The variable stars called Cepheids have an average luminosity (absolute brightness) that is strongly correlated with the period over which the Cepheids' light variation occurs. More luminous Cepheids take longer for each cycle of variation. *Drawing by Crystal Stevenson*

pinwheel of stars. We view the galaxy edgewise from a point near the outside of one of the spirals.

Once astronomers had learned how to estimate distances, they applied their method throughout the Milky Way and beyond. They could then correlate the distances to the stars with stellar motions, derived from the Doppler effect (see Figure 28). The spectral lines in starlight have revealed the stars' motions to us. Most of the stars in a particular region of our galaxy have roughly the same speed and the same direction of motion. This allows the deduction that the vast pinwheel of the Milky Way is rotating, as each star orbits the galactic center.

But the details of these orbits have long puzzled astronomers. As they measured the orbital speeds of stars at distances farther and farther from the center of the galactic pinwheel, the speeds of stars farther from the center than the sun did not decrease. This rotation curve (see Figure 29) does not correspond to what astronomers expect from gravity and its effects on orbital motion.

Just as the planets farther from the sun move more slowly in their orbits, stars farther from the galactic center should be orbiting more slowly. This should be true be-

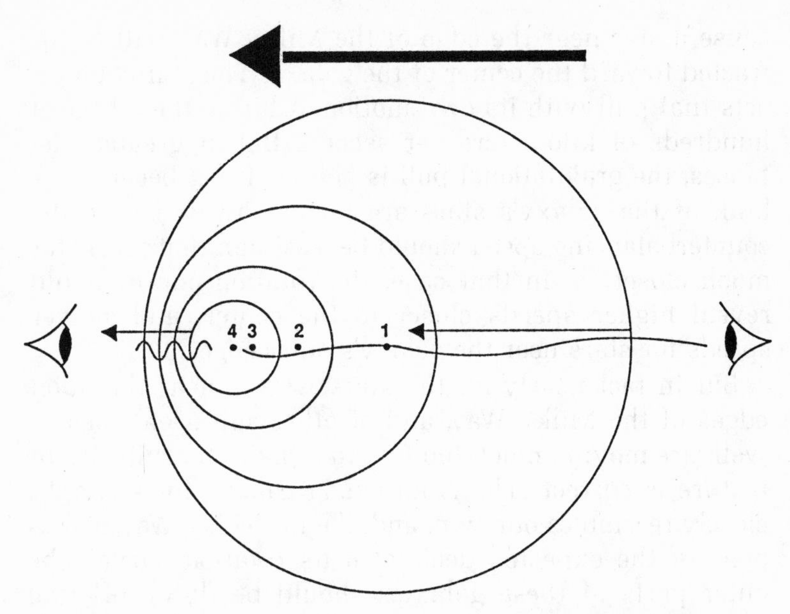


Figure 28: The Doppler effect arises from the relative motion of a source and an observer. This effect changes all the wavelengths and frequencies by the same fractional amount. This fraction increases as the relative velocity of source and observer increases. Recessional motions decrease the frequencies and increase the wavelengths, whereas motions of approach increase the frequencies and decrease the wavelengths. *Drawing by Crystal Stevenson*

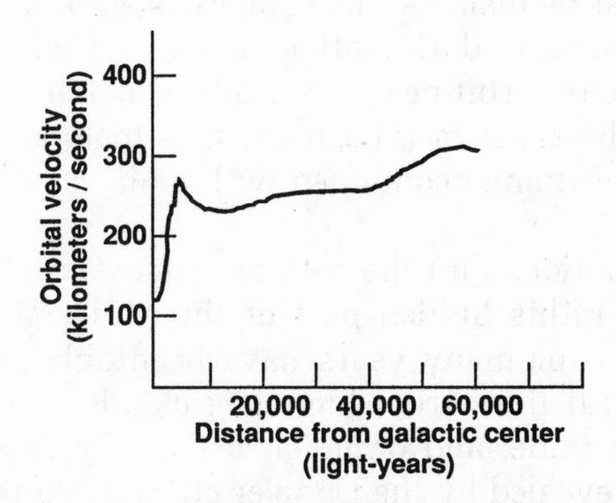


Figure 29: A galaxy's rotation curve shows the orbital velocities measured at different distances from the galaxy's center. Most spiral galaxies have rotation curves with less of a decrease in velocity at large distances from the center than would be expected from the observed distribution of visible matter. *Drawing by Crystal Stevenson*

cause a star near the edge of the Milky Way will be attracted toward the center of the galaxy. The star counteracts that pull with its own motion, orbiting the center at hundreds of kilometers per second. But at greater distances, the gravitational pull is not so strong because the bulk of the galaxy's stars are farther away. Hence the counterbalancing speed should be less than that for a star much closer in. In that case, the rotation curve should reveal higher speeds closer to the center, and slower speeds for stars near the galaxy's outer edges.

But in fact nearly all the stars we see near the outer edges of the Milky Way, and of other spiral galaxies as well, are moving much too fast to remain in orbit—if our picture is correct. The Andromeda galaxy, for example, closely resembles our own, and like the Milky Way, shows none of the expected decline in its rotation curve. The outer parts of these galaxies should be flying off into space. Yet the galaxies have existed for ten billion years or more.

Gravity provides the answer—but only if the Milky Way and its cousins contain far more mass than we see in stars. In order to explain rotation curves like that shown in Figure 29, astronomers conclude that the visible disk of the galaxy must be inside a huge sphere of dark matter. The additional mass of dark matter can explain why stars far from the center orbit nearly as rapidly as those closer in. Whatever this mass may be, it amounts to about *ten times more matter* than is contained within all the stars in the galaxy.

Stellar motions and the rotation curve aren't the only evidence for this hidden part of the Milky Way. Radio astronomers for many years have been able to observe radio spectral lines from huge gas clouds of hydrogen, carbon monoxide, and other molecules. By tracking the velocities, revealed by the Doppler effect of these tenuous but ubiquitous clouds, the pinwheel's details become not only clearer but also more suggestive.

The galaxy's rotation curve, as revealed by the gas

clouds, shows "flatness" at all the velocities at which stellar motions exhibit a similarly flat rotation curve. But in addition, certain clouds of hydrogen believed to lie at the periphery of the pinwheel are moving at especially great speeds. For these clouds to remain gravitationally attached to the galaxy, a huge amount of dark matter must exist.

In addition to observational evidence, theoretical arguments also imply enormous amounts of dark matter. Without such dark matter, calculations show, the galactic pinwheel, instead of remaining flat, would warp, buckle, and eventually break apart, so that the galaxy would have long since lost its disklike structure.

All these lines of evidence imply that the Milky Way contains far more dark matter than visible matter. The best estimates are that the Milky Way and similar galaxies comprise ten times more dark matter than visible matter. The estimates also imply that most of the dark matter must be distributed in a roughly spherical shape at distances far beyond the boundaries of the visible Milky Way.

## WHAT IS THE DARK MATTER?

With dark matter accepted as real, the question quickly becomes one of the *nature* of the dark matter. We can easily eliminate certain possibilities. The first of these includes any type of stars (see Table 2). Virtually any fusion-fired star in the galaxy, or any amount of gas, can be shown theoretically to emit enough radiation for us to detect. Even if we had problems in detecting one feeble star, the billions of subluminous stars needed to explain the vast quantities of dark matter in the Milky Way would surely be observable as a glow in the night sky.

But variants of stars offer a family of dark matter candidates. These range from the brown dwarfs (see Chapter 11) to black holes to white dwarfs. We may examine each of these in turn.

Brown dwarfs seem a likely possibility. But since each

| DARK MATTER CANDIDATES | |
|---|---|
| Stars | Black holes<br>White dwarfs<br>Brown dwarfs |
| Macroparticles | "Jupiters"<br>"Cosmic rocks" |
| Microparticles | WIMPs<br>Neutrinos<br>Axions<br>Others |

Table 2: The chief candidates for dark matter include hypothetical elementary particles such as WIMPs and axions; known elementary particles called neutrinos; rocks and planetlike objects; and brown dwarfs.

has so little mass (less than a hundredth the mass of the sun) astronomers could require enormous numbers of them to provide the dark matter. As discussed in Chapter 11, we have yet to find a single definitive example of these brown dwarfs in our own cosmic neighborhood. This makes brown dwarfs a poor candidate.

Black holes are, as might be guessed, the ultimate convenient explanation for almost every astrophysical puzzle. They may nevertheless prove to be the dark matter—if the Milky Way turns out to contain about a million black holes of a million solar masses, or a trillion black holes with one solar mass each. But we have no idea how and why black holes could be made in such numbers. We still have no compelling explanation of how a star, or a group of stars, could become a million-solar-mass black hole. The mass might well explode long before imploding to produce such a gargantuan, albeit invisible, star. Furthermore we would expect to see prodigious amounts of ultraviolet radiation from the accretion disks that would form around these black holes. Hence black holes have their difficulties in explaining this dark matter.

White dwarfs are among the most promising candidates. Recall that these are burnt-out stellar cores, each with about the mass of the sun. Computer models tell us that white dwarfs, already feeble sources of light, will burn out completely after a few tens of billions of years. Still, a trillion burnt-out white dwarfs would be needed to furnish the dark matter in the Milky Way. Why did so many white dwarfs burn out? Why aren't enough still shining brightly enough for us to see them? Astronomers have no good answers to these questions.

Brown dwarfs, black holes, and burnt-out white dwarfs provide what we may call the stellar possibilities for the dark matter. Because each of these has its difficulties, astronomers also consider a different set of possibilities, more exotic and thus more intriguing. This set consists of hypothetical new types of elementary particles, particles that are hypothesized to interact only rarely with "ordinary" matter, and they emit little or no light or other types of radiation. Thanks to the inventiveness of particle theorists, the hypothetical particles have fascinating names such as para-photons, higgsinos, majorons, preons, shadow matter, and cosmic strings. All of these are predicted to exist by various theories of particle physics such as the class called grand unified theories (GUTs). They are all tiny subparts of atoms, fractions of an electron's size, and are wonderfully, almost capriciously, elusive. We don't see them here, but then again, only recently have we decided it's worth looking for them. If one of these particle types is to explain the dark matter, it must not merely *exist* but must be the most abundant type of particle in the universe!

Where would these particles have come from? Most theories predict that they would have been made at the beginning of the universe and would have lifetimes of many billions of years. Since the particles produce no radiation, they would be impossible to detect in light. And since they don't agglomerate in large masses like stars, their individual gravitational effects are infinitesimal. Only through the effect of their sheer numbers, spread through the dark of space, would they provide the gravitational pull that the dark matter reveals through its effects on stellar motions. The hypothetical elementary particles would provide a strange dark matter picture—a galaxy that is predominantly a vast collection of tiny invisible particles.

Astronomers now eagerly await news from particle-physics laboratories. Exotic particles may provide the answer to the mystery of dark matter—if they turn out to exist. The physics laboratory has therefore become the dark matter–particle test bed. Here on Earth, we can hunt for these particles by looking for the extremely rare occasions when these particles would interact with ordinary matter, either producing radiation or making new elements from old. Some exotic particle types might be created artificially by making other common particle types collide in an accelerator. Existence, if proved, does not automatically imply that any of these particle types will account for the dark matter. But without a demonstration of their existence, using them to explain dark matter may seem even more bizarre than the existence of the dark matter itself. As you read this, particle physicists labor at gargantuan accelerators, seeking the fleeting evidence that one or more of these dark matter candidates does indeed exist. It will be big news in the worlds of physics and astronomy if they ever find them.

## GOODBYE TO GRAVITY?

The claim that nine-tenths of our galaxy consists of dark matter has led to the most unusual but probably the least likely suggestion of how to explain this result: Dark matter doesn't exist, and the laws of gravity are different from what we believe! This theory has been proposed by the physicist Mordecai Milgrom, who suggests that our current understanding of gravity is accurate for relatively modest distances, but for great distances must be modified. Milgrom's theory would mean that the Milky Way's flat rotation curve arises not from huge amounts of unseen matter but because gravity does not follow the generally accepted rules. The implications of Milgrom's idea are intriguing for the galaxy and for the universe as a whole: Stars would be bound by different scales of forces than previously imagined, time would likewise be different, mass would change in unfamiliar ways, and so on.

This sounds like science fiction. And it's hard to prove right or wrong because Milgrom's theory requires that gravity be different from what we think only over immense distance scales. Earth-based experiments would be unaffected. So too would solar-system studies. It's an interesting idea, but it offers no proposed observation that would distinguish it as a unique explanation. Milgrom's hypothesis could explain the dark matter, but so might many other possibilities. The idea that gravity must be modified will receive serious consideration only when more compelling evidence is found. For now, gravity is the last theory that physicists and astronomers are willing to modify or abandon.

We have reached the end of our survey of the mysteries of the Milky Way—without exhausting the mysteries themselves. Some of the former puzzles have now resolved themselves into fairly well-understood chains of cosmic evolution, such as those involved in the formation, maturity, and death of stars. Others, such as the possibilities of other planets or the nature of the dark matter that rules the universe, remain almost as intractable today as they were when humans first perceived them as important topics for investigation.

Curiosity, the emotional force that has made humanity what we are today, will continue to drive astronomers onward in the search to resolve today's mysteries and thus to confront tomorrow's. Through times of great patience and discovery, through lean and full funding, the men and women who are drawn to examine the universe will do whatever they can to find, to attack, and to probe its secrets successfully. As they persevere, they know that unanswered questions will often yield to determined investigation. Of course, new questions continually emerge as old ones are answered. Those who share in scientists' hopes for understanding the cosmos eagerly await the day when astronomers resolve, at least in part, the current mysteries of the Milky Way.

# INDEX